Adaptive Machine Learning Algorithms with Python

Solve Data Analytics and Machine Learning Problems on Edge Devices

Chanchal Chatterjee

Apress®

Adaptive Machine Learning Algorithms with Python: Solve Data Analytics and Machine Learning Problems on Edge Devices

Chanchal Chatterjee
San Jose, CA, USA

ISBN-13 (pbk): 978-1-4842-8016-4   ISBN-13 (electronic): 978-1-4842-8017-1
https://doi.org/10.1007/978-1-4842-8017-1

Managing Director, Apress Media LLC: Welmoed Spahr
Acquisitions Editor: Celestin Suresh John
Development Editor: James Markham
Coordinating Editor: Mark Powers
Copy Editor: Mary Behr

Cover designed by eStudioCalamar

Cover image by Shubham Dhage on Unsplash (www.unsplash.com)

Distributed to the book trade worldwide by Apress Media, LLC, 1 New York Plaza, New York, NY 10004, U.S.A. Phone 1-800-SPRINGER, fax (201) 348-4505, e-mail orders-ny@springer-sbm.com, or visit www.springeronline.com. Apress Media, LLC is a California LLC and the sole member (owner) is Springer Science + Business Media Finance Inc (SSBM Finance Inc). SSBM Finance Inc is a **Delaware** corporation.

For information on translations, please e-mail booktranslations@springernature.com; for reprint, paperback, or audio rights, please e-mail bookpermissions@springernature.com.

Apress titles may be purchased in bulk for academic, corporate, or promotional use. eBook versions and licenses are also available for most titles. For more information, reference our Print and eBook Bulk Sales web page at www.apress.com/bulk-sales.

Any source code or other supplementary material referenced by the author in this book is available to readers on GitHub (https://github.com/Apress). For more detailed information, please visit www.apress.com/source-code.

Printed on acid-free paper

I dedicate this book to my father, Basudev Chatterjee, and all my teachers and mentors who have guided and inspired me.

Table of Contents

About the Author

Chanchal Chatterjee, Ph.D., has held several leadership roles in machine learning, deep learning, and real-time analytics. He is currently leading machine learning and artificial intelligence at Google Cloud Platform, California, USA. Previously, he was the Chief Architect of EMC CTO Office where he led end-to-end deep learning and machine learning solutions for data centers, smart buildings, and smart manufacturing for leading customers. Chanchal has received several awards including an Outstanding Paper Award from the IEEE Neural Network Council for adaptive learning algorithms, recommended by MIT professor Marvin Minsky. Chanchal founded two tech startups between 2008-2013. Chanchal has 29 granted or pending patents and over 30 publications. Chanchal received M.S. and Ph.D. degrees in Electrical and Computer Engineering from Purdue University.

About the Technical Reviewer

Joos Korstanje is a data scientist with over five years of industry experience in developing machine learning tools, a large part of which are forecasting models. He currently works at Disneyland Paris where he develops machine learning for a variety of tools.

Acknowledgments

I want to thank my professor and mentor Vwani Roychowdhury for guiding me through my Ph.D. thesis, where I first created much of the research presented in this book. Vwani taught me how to research, write, and present this material in the many papers we wrote together. He also inspired me to continue this research and eventually write this book. I could not have done it without his inspiration, help, and guidance. I sincerely thank Vwani for being my teacher and mentor.

Preface

This book presents several categories of streaming data problems that have significant value in machine learning, data visualization, and data analytics. It offers many adaptive algorithms for solving these problems on streaming data vectors or matrices. Complex neural network-based applications are commonplace and computing power is growing exponentially, so why do we need adaptive computation?

Adaptive algorithms are critical in environments where the data volume is large, data has high dimensions, data is time-varying and has changing underlying statistics, and we do not have sufficient storage, computing power, and bandwidth to process the data with low latency. One such environment is computation on edge devices.

Due to the rapid proliferation of billions of devices at the cellular edge and the exponential growth of machine learning and data analytics applications on these devices, there is an urgent need to manage the following on these devices:

- Power usage for computation at scale

- Non-stationarity of inputs and drift of the incoming data

- Latency of computation on devices

- Memories and bandwidth of devices

The 2021 Gartner report on Edge computation [Stratus Technologies, 2021] suggests that device-based computation growth propelled by the adoption of cloud and 5G will require us to prioritize and build a framework for edge computation.

In these environments are the following constraints:

- The data cannot be batched immediately and needs to be used instantaneously. We have a streaming sequence of vectors or matrices as inputs to the algorithms.

- The data changes with time. In other words, the data is non-stationary, causing significant drift of input features whereby the machine learning models are no longer effective over time.

- The data volume and dimensionality are large, and we do not have the device memory, bandwidth, or power to store or upload the data to the cloud.

In these applications, we use adaptive algorithms to manage the device's power, memory, and bandwidth so that we can maintain accuracy of the pretrained models. Some use cases are the following:

1. Calculate feature drift of incoming data and detect training-serving skew [Kaz et al. 2021] ahead of time.

2. Adapt to incoming data drift and calculate features that best fit the data.

3. Calculate anomalies in incoming data so that good, clean data is used by the models.

4. Compress incoming data into features for use in new model creation.

In Chapter 8, I present solutions to these problems with adaptive algorithms discussed with real-world data.

Detecting Feature Drift

See the example where the real-time data [Vinicius Souza et al. 2020] has a gradual drift of features. In real-time environments there are changes in the underlying distribution of the observed data. These changes in the statistical properties of the data are called *drift*. When the changes in statistical properties are smooth, it is called *gradual drift*. Figure 1 shows a slow change in the magnitude of the components of a multivariate data over time, showing a gradual drift.

It's important to detect these drift components early in the process so that we can update the machine learning model to maintain performance. Note that the baseline statistics of the data is not known ahead of time.

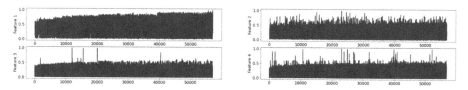

Figure 1. *Data components show gradual drift over time*

I used an adaptive algorithm (see Chapter 5) to compute the principal components [principal component analysis, Wikipedia] of the data and derive a metric from them. Figure 2 shows that the metric does not converge to its statistical value (ideally 1) and diverges towards 0. This detects the feature drift quickly so that the edge device can update the machine learning model.

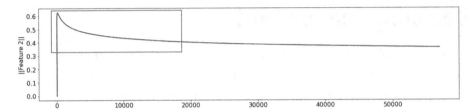

Figure 2. *An adaptive principal component-based metric detects drift in data early in the real-time process*

The downward slope of the detection metric in the graph indicates the gradual drift of the features.

Adapting to Drift

In another type of drift, the data changes its statistical properties abruptly. Figure 3 shows simulated multi-dimensional data that abruptly changes to a different underlying statistic after 500 samples.

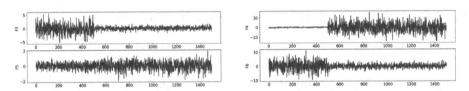

Figure 3. *Simulated data that abruptly changes statistical properties after 500 samples*

The adaptive algorithms help us adapt to this abrupt change and recalculate the underlying statistics, in this case, the first two principal eigenvectors of the data (see Chapter 6). The ideal values in Figure 4 are 1. As the data changes abruptly after 500 samples, the value falls and quickly recovers back to 1.

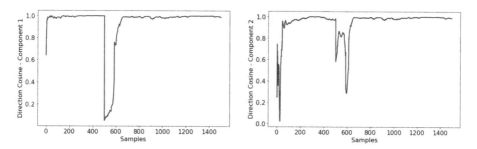

Figure 4. *Principal eigenvectors adapt to abruptly changing data*

My Approach

Adaptive Algorithms and Best Practices

In this book, I offer over 50 examples of adaptive algorithms to solve real-world problems. I also offer best practices to select the right algorithms for different use cases. I formulate these problems as matrix computations, where the underlying matrices are unknown. I assume that the entire data is not available at once. Instead, I have a sequence of random matrices or vectors from which I compute the matrix functions without knowing the matrices. The matrix functions are computed adaptively as each sample is presented to the algorithms.

1. My algorithms process each incoming sample $\mathbf{x}_k$ such that at any instant k all of the currently available data is taken into consideration.

2. For each sample, the algorithm estimates the desired matrix functions, and its estimates have known statistical properties.

3. The algorithms have the ability to adapt to variations in the underlying systems so that for each sample $\mathbf{x}_k$ I obtain the current state of the process and not an average over all samples (i.e., I can handle both stationary and non-stationary data).

Problems with Conventional Non-Adaptive Approaches

The conventional approach for evaluating matrix functions requires the computation of the matrix after collecting all samples and then the application of a numerical procedure. There are two problems with this approach.

1. The dimension of the samples may be large so that even if all the samples are available, performing the matrix algebra may be difficult or may take a prohibitively large number of computational resources and memory.

2. The matrix functions evaluated by conventional schemes cannot adapt to small changes in the data (e.g., a few incoming samples). If the matrix functions are estimated by conventional methods from K (finite) samples, then for each additional sample, all of the computation must be repeated.

These deficiencies make the conventional schemes inefficient for real-time applications.

Computationally Simple

My approach is to use computationally simple adaptive algorithms. For example, given a sequence of random vectors $\{\mathbf{x}_k\}$, a well-known algorithm for the principal eigenvector evaluation uses the update rule $\mathbf{w}_{k+1} = \mathbf{w}_k + \eta$ $(\mathbf{x}_k\mathbf{x}_k^T - \mathbf{w}_k\mathbf{w}_k^T\mathbf{x}_k\mathbf{x}_k^T\mathbf{w}_k)$, where η is a small positive constant. In this algorithm, for each sample $\mathbf{x}_k$ the update procedure requires simple matrix-vector multiplications, yet the vector $\mathbf{w}_k$ converges quickly to the principal eigenvector of the data correlation matrix. Clearly, this can be easily implemented in CPUs on devices with low memory and power usage.

The objective of this book is to present a variety of neuromorphic [neuromorphic engineering, Wikipedia] adaptive algorithms [Wikipedia] for matrix algebra problems. Neuromorphic algorithms work by mimicking the physics of the human neural systems using networks, where activation by neurons propagate to other neurons in a cascading chain.

Matrix Functions I Solve

The matrix algebra functions that I compute adaptively include

- Normalized mean, median

- LU decomposition (square root), inverse square root

- Eigenvector/eigenvalue decomposition (EVD)

- Generalized EVD

- Singular value decomposition (SVD)

- Generalized SVD

For each matrix function, I will discuss practical use cases in machine learning and data analytics and support them with experimental results.

Common Methodology to Derive Adaptive Algorithms

Another contribution of this book is the presentation of a **common methodology** to derive each adaptive algorithm. For each matrix function and every adaptive algorithm, I present a scalar unconstrained objective function $J(W;A)$ whose minimizer W^* is the desired matrix function of A. From this objective function $J(W;A)$, I derive the adaptive algorithm such as $W_{k+1} = W_k - \eta \nabla_w J(W_k; A)$ by using standard techniques of optimization (for example, gradient descent). I then speed up these adaptive algorithms by using statistical methods. Note that this helps practitioners create new adaptive algorithms for their use cases.

In summary, the book starts with a common framework to derive adaptive algorithms and then uses the framework for each category of streaming data problems starting with the adaptive median to complex principal components and discriminant analysis. I present practical problems in each category and derive the algorithms to solve them. The final chapter solves critical edge computation problems for time-varying, non-stationary data with minimal compute, memory, latency, and bandwidth. I also provide the code [Chatterjee GitHub] for all algorithms discussed here.

GitHub

All simulations and implementation code by chapters are published in the public GitHub:

https://github.com/cchatterj0/AdaptiveMLAlgorithms

The GitHub page contains the following items:

- Python code for all chapters starting with Chapter 2

- MATLAB simulation code in a separate directory

- Data for all implementation code

- Proofs of convergence for some of the algorithms

CHAPTER 1

Introduction

In this chapter, I present the adaptive computation of important features for data representation and classification. I demonstrate the importance of these features in machine learning, data visualization, and data analytics. I also show the importance of these algorithms in multiple disciplines and present how these algorithms are obtained there. Finally, I present a common methodology to derive these algorithms. This methodology is of high practical value since practitioners can use this methodology to derive their own features and algorithms for their own use cases.

For these data features, I assume that the data arrives as a sequence, has to be used instantaneously, and the entire batch of data cannot be stored in memory.

In machine learning and data analysis problems such as regression, classification, enhancement, or visualization, effective representation of data is key. When this data is multi-dimensional and time varying, the computational challenges are more formidable. Here we not only need to compute the represented data in a timely manner, but also adapt to the changing input in a fast, efficient, and robust manner.

A well-known method of data compression/representation is the Karhunen-Loeve theorem [Karhunen–Loève theorem, Wikipedia] or eigenvector orthonormal expansion [Fukunaga 90]. This method is also known as principal component analysis (PCA) [principal component analysis, Wikipedia]. Since each eigenvector can be ranked by its

© Chanchal Chatterjee 2022
C. Chatterjee, *Adaptive Machine Learning Algorithms with Python*,
https://doi.org/10.1007/978-1-4842-8017-1_1

corresponding eigenvalue, a subset of the "best" eigenvectors can be chosen as the most relevant features.

Figure 1-1 shows two-class, two-dimensional data in which the best feature for data representation is the projection of the data on vector e_1, which captures the most significant properties of the data from the two classes.

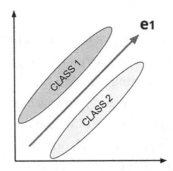

Figure 1-1. *Data representation feature e_1 that best represents the data from the two classes [Source: Chatterjee et al. IEEE Transactions on Neural Networks, Vol. 8, No. 3, pp. 663-678, May 1997]*

In classification, however, you generally want to extract features that are effective for preserving *class separability*. Simply stated, the goal of classification feature extraction is to find a transform that maps the raw measurements into a smaller set of features, which contain all the discriminatory information needed to solve the overall pattern recognition problem.

Figure 1-2 shows the same two-class, two-dimensional data in which the best feature for classification is vector e_2, whereby projection of the data on e_2 leads to best class separability.

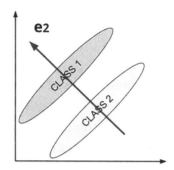

Figure 1-2. *Data classification feature e_2 that leads to best class separability [Source: Chatterjee et al. IEEE Transactions on Neural Networks, Vol. 8, No. 3, pp. 663-678, May 1997]*

Before I present the adaptive algorithms for streaming data in the following chapters, I need to discuss the following:

1. Commonly used features that are obtained by a linear transform of the data and are used with streaming data for edge applications

2. Historical relevance of these features and how we derive them from different disciplines

3. Why we want to use adaptive algorithms to compute these features

4. How to create a common mathematical framework to derive adaptive algorithms for these features and many more

1.1 Commonly Used Features Obtained by Linear Transform

In this section, I discuss four commonly used features for data analytics and machine learning. These features are effective in data classification and representation, and can be easily obtained by a simple linear transform of the data. The simplicity and effectiveness of these features makes them useful for streaming data and edge applications.

In mathematical terms, let $\{\mathbf{x}_k\}$ be an n-dimensional (zero mean) sequence that represents the data. We are seeking a matrix sequence $\{W_k\}$ and a transform:

$$\mathbf{y}_k = W_k^T \mathbf{x}_k, \tag{1.1}$$

such that the linear transform $\mathbf{y}_k$ has properties of data representation and is our desirable feature. I discuss a few of these features later.

Definition: Define the data correlation matrix A of $\{\mathbf{x}_k\}$ as

$$A = \lim_{k \to \infty} E\left[\mathbf{x}_k \mathbf{x}_k^T \right]. \tag{1.2}$$

Data Whitening

Data whitening is a process of decorrelating the data such that all components have unit variance. It is a data preprocessing step in machine learning and data analysis to "normalize" the data so that it is easier to model. Here the linear transform $\mathbf{y}_k$ of the data has the property $E\left[\mathbf{y}_k \mathbf{y}_k^T \right] = I_n$ (identity). I discuss in Chapter 3 that the optimal value of $W_k = A^{-\frac{1}{2}}$.

Figure 1-3 shows the correlation matrices of the original and whitened data. The original random normal data is highly correlated as shown by the colors on all axes. The whitened data is fully uncorrelated with no correlation between components since only diagonal values exist.

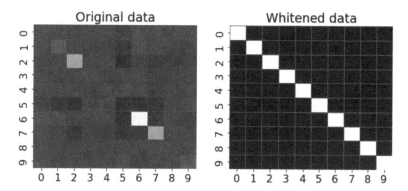

Figure 1-3. *Original correlated data is uncorrelated by the data whitening algorithm*

The Python code to generate the whitened data from original dataset X[nDim, nSamples] is

```python
from scipy.linalg import eigh
# Compute Correlation matrix and eigen vectors of the
original data
corX = (X @ X.T) / nSamples
# generate the whitened data
eigvals, eigvecs = eigh(corX)
V = np.fliplr(eigvecs)
D  = np.diag(np.sqrt(1/eigvals[::-1]))
Y = V @ D @ V.T @ X
corY = (Y @ Y.T)/nSamples
# Plot the original and whitened correlation matrices
import seaborn as sns
plt.figure(figsize=(10, 4))
plt.rcParams.update({'font.size': 16})
plt.subplot(1, 2, 1)
sns.heatmap(corX, linewidth=0.5, linecolor="green",
cmap='RdBu', cbar=False)
```

```
plt.title("Original data")
plt.subplot(1, 2, 2)
sns.heatmap(corY, linewidth=0.5, linecolor="green", cmap='hot',
cbar=False)
plt.title("Whitened data")
plt.show()
```

Principal Components

Principal component analysis (PCA) is a well-studied example of the data representation model. From the perspective of *classical statistics*, PCA is an analysis of the covariance structure of multivariate data $\{\mathbf{x}_k\}$. Let $\mathbf{y}_k=[y_{k1},...,y_{kp}]^T$ be the components of the PCA-transformed data. In this representation, the first principal component y_{k1} is a one-dimensional linear subspace where the variance of the projected data is maximal. The second principal component y_{k2} is the direction of maximal variance in the space orthogonal to the y_{k1} and so on.

It has been shown that the optimal weight matrix W_k is the eigenvector matrix of the correlation of the zero-mean input process $\{\mathbf{x}_k\}$. Let $A\Phi=\Phi\Lambda$ be the *eigenvector decomposition (EVD)* of A, where Φ and Λ are respectively the eigenvector and eigenvalue matrices. Here $\Lambda=diag(\lambda_1,...,\lambda_n)$ is the diagonal eigenvalue matrix with $\lambda_1\geq...\geq\lambda_n>0$ and Φ is orthonormal. We denote $\Phi_p\in\mathfrak{R}^{n\times p}$ as the matrix whose columns are the first p principal eigenvectors. Then optimal $W_k=\Phi_p$.

There are three variations of PCA that are useful in applications.

1. When $p=n$, the n components of $\mathbf{y}_k$ are ordered according to maximal to minimal variance. This is the component analyzer that is used for *data analysis*.

2. When $p<n$, the p components of $\mathbf{y}_k$ have maximal information for *data compression*.

3. For $p<n$, the $n-p$ components of $\mathbf{y}_k$ with minimal
    variance can be regarded as abnormal signals
    and reconstructed as $(I_n-\Phi_p\Phi_p{}^T)\mathbf{x}_k$ to obtain a
    novelty filter.

Figure 1-4 shows the correlation matrices of the original and PCA-transformed random normal data. The original data is highly correlated as shown by the colors on all axes. The PCA-transformed data is uncorrelated with diagonal blocks only and highest representation value for component 1 (top left corner block) and decreasing thereafter.

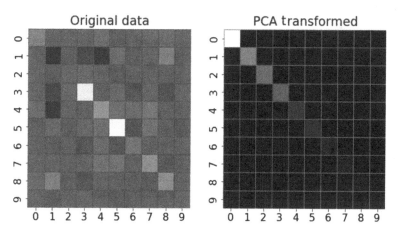

Figure 1-4. *Original correlated data is uncorrelated by PCA projection*

The Python code for the PCA projected data from original dataset X[nDim, nSamples] is

```python
from scipy.linalg import eigh
# Compute data correlation matrix
corX = (X @ X.T) / nSamples
# generate the PCA transformed data
eigvals, eigvecs = eigh(corX)
EstV = np.fliplr(eigvecs)
```

```
Y = EstV.T @ X
corY = (Y @ Y.T)/nSamples
# plot the PCA transformed data
import seaborn as sns
plt.figure(figsize=(10, 5))
plt.rcParams.update({'font.size': 16})
plt.subplot(1, 2, 1)
sns.heatmap(corX, linewidth=0.5, linecolor="green",
cmap='RdBu', cbar=False)
plt.title("Original data")
plt.subplot(1, 2, 2)
sns.heatmap(corY, linewidth=0.5, linecolor="green", cmap='hot',
cbar=False)
plt.title("PCA Transformed")
plt.show()
```

Linear Discriminant Features

Linear discriminant analysis (LDA) [linear discriminant analysis, Wikipedia] creates features from data from multiple classes such that the transformed representation $\mathbf{y}_k$ has the most class separability. Given a data $\mathbf{x}_k$ and the corresponding classes $\mathbf{c}_k$, we can calculate the following matrices:

- data correlation matrix $B=E(\mathbf{x}_k\mathbf{x}_k^T)$,

- cross correlation matrix $M=E(\mathbf{x}_k\mathbf{c}_k^T)$,

- and $A=MM^T$.

It is well known that the linear transform W_k (Eq 1.1) is the *generalized eigen-decomposition* (GEVD) [generalized eigenvector, Wikipedia] of A with respect to B. Here $A\Psi=B\Psi\Delta$ where Ψ and Δ are respectively the generalized eigenvector and eigenvalue matrices. Furthermore, $\Psi_p\in\Re^{n\times p}$ is the matrix whose columns are the first $p\leq n$ principal generalized eigenvectors.

Figure 1-5 shows a two-class classification problem where the real-world data correlation matrix on the left has values on all axes and it is hard to distinguish the two classes. On the right is the LDA-transformed data, which clearly shows the two classes and is easy to classify.

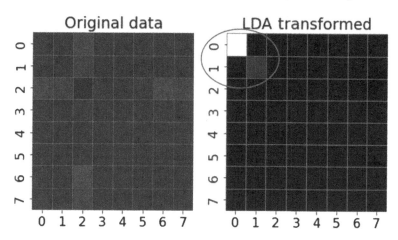

Figure 1-5. *Original correlated data is uncorrelated by linear discriminant analysis*

The Python code to adaptively generate the LDA transformed correlation matrix from a two-class, multi-dimensional dataset [nDim, nSamples] is

```
# Adaptively compute matrices A and B
from numpy import linalg as la
dataset2 = dataset.drop(['Class'],1)
nSamples = dataset2.shape[0]
nDim = dataset2.shape[1]
classes = np.array(dataset['Class']-1)
classes_categorical = tf.keras.utils.to_categorical(classes,
num_classes=2)
M = np.zeros(shape=(nDim,2)) # stores adaptive
correlation matrix
B = np.zeros(shape=(nDim,nDim)) # stores adaptive
correlation matrix
```

```
for iter in range(nSamples):
    cnt = iter + 1
    # generate matrices A and B from current sample x
    x = np.array(dataset2.iloc[iter])
    x = x.reshape(nDim,1)
    B = B + (1.0/cnt)*((np.dot(x, x.T)) - B)
    y = classes_categorical[iter].reshape(2,1)
    M = M + (1.0/cnt)*((np.dot(x, y.T)) - M)
    A = M @ M.T
# generate the LDA transformed data
from scipy.linalg import eigh
from sklearn.preprocessing import normalize
eigvals, eigvecs = eigh(A, B)
V = np.fliplr(eigvecs)
VTAV = np.around(V.T @ A @ V, 2)
VTBV = np.around(V.T @ B @ V, 2)
# plot the LDA transformed data
import seaborn as sns
plt.figure(figsize=(8, 8))
plt.rcParams.update({'font.size': 16})
plt.subplot(2, 2, 1)
sns.heatmap(A, linewidth=0.5, linecolor="green", cmap='RdBu',
cbar=False)
plt.title("Original data")
plt.subplot(2, 2, 2)
sns.heatmap(VTBV, linewidth=0.5, linecolor="green", cmap='hot',
cbar=False)
plt.title("LDA Transformed")
plt.subplot(2, 2, 3)
sns.heatmap(A, linewidth=0.5, linecolor="green", cmap='RdBu',
cbar=False)
plt.subplot(2, 2, 4)
```

```
sns.heatmap(VTAV, linewidth=0.5, linecolor="green", cmap='hot',
cbar=False)
plt.show()
```

Singular Value Features

Singular value decomposition (SVD) [singular value decomposition, Wikipedia] is a special case of a EVD problem as follows. Given the cross-correlation (n-by-m real) matrix $M=E(\mathbf{x}_k\mathbf{c}_k^T) \in \Re^{nXm}$, SVD computes two matrices $U_k \in \Re^{nXn}$ and $V_k \in \Re^{mXm}$ such that $U_k^T M V_k = S_{nxm}$, where U_k and V_k are orthonormal and $S=\text{diag}(s_1,...s_r)$, $r=\min(m, n)$, with $s_1>=...>=s_r>=0$. By rearranging the vectors $\mathbf{x}_k$ and $\mathbf{c}_k$ we can make a nxm dimensional SVD problem into a $(n+m)x(n+m)$ dimensional EVD problem.

Summary

Table 1-1 summarizes the discussion in this section. Note that given a sequence of vectors $\{\mathbf{x}_k\}$, we are seeking a matrix sequence $\{W_k\}$ and a linear transform $\mathbf{y}_k = W_k^T \mathbf{x}_k$.

Table 1-1. *Statistical Property of $\mathbf{y}_k$ and Matrix Property of W_k*

Computation	Statistical Property of $\mathbf{y}_k$	Matrix Property of W_k
Whitening	$E\left[\mathbf{y}_k\mathbf{y}_k^T\right]=I_n$	$W_k = A^{-\frac{1}{2}} = \Phi\Lambda^{-\frac{1}{2}\Phi T}$
PCA/EVD	Max $E\left[\mathbf{y}_k^T\mathbf{y}_k\right]$ subj. to $W_k^T W_k = I_p$	$W_k = \Phi_p$ and $E\left[\mathbf{y}_k\mathbf{y}_k^T\right]=I_p$
LDA/GEVD	Max $E\left[\mathbf{y}_k^T\mathbf{y}_k\right]$ subj. to $W_k^T B W_k = I_p$	$W_k = \Psi_p$ and $E\left[\mathbf{y}_k\mathbf{y}_k^T\right]=I_p$

1.2 Multi-Disciplinary Origin of Linear Features

In this section, I further discuss the importance of data representation and classification features by showing how multiple disciplines derive these features and compute them adaptively for streaming data. For each discipline, I demonstrate the use of these features on real data.

Hebbian Learning or Neural Biology

Hebb's postulate of learning is the oldest and most famous of all learning rules. Hebb proposed that when an axon of cell A excites cell B, the synaptic weight W is adjusted based on $f(\mathbf{x}, \mathbf{y})$, where $f(\cdot)$ is a function of presynaptic activity $\mathbf{x}$ and postsynaptic activity $\mathbf{y}$. As a special case, we may write the weight adjustment $\Delta W = \eta \mathbf{x} \mathbf{y}^T$, where $\eta > 0$ is a small constant.

Given an input sequence $\{\mathbf{x}_k\}$, we can construct its linear transform $y_k = \mathbf{w}_k^T \mathbf{x}_k$, where $\mathbf{w}_k \in \Re^n$ is the weight vector. We can describe a Hebbian rule [Haykin 94] for adjusting $\mathbf{w}_k$ as $\mathbf{w}_{k+1} = \mathbf{w}_k + \eta \mathbf{x}_k y_k$. Since this rule leads to exponential growth, Oja [Oja 89] introduced a rule to limit the growth of $\mathbf{w}_k$ by normalizing $\mathbf{w}_k$ as follows:

$$\mathbf{w}_{k+1} = \frac{\mathbf{w}_k + \eta \mathbf{x}_k y_k}{\|\mathbf{w}_k + \eta \mathbf{x}_k y_k\|}, \tag{1.3}$$

Assuming small η and $\|\mathbf{w}_k\| = 1$, (1.3) can be expanded as a power series in η, yielding

$$\mathbf{w}_{k+1} = \mathbf{w}_k + \eta \left(\mathbf{x}_k y_k - y_k^2 \mathbf{w}_k \right) + O\left(\eta^2 \right). \tag{1.4}$$

Eliminating the $O(\eta^2)$ term, we get the *constrained Hebbian algorithm* (*CHA*) or *Oja's one-unit rule* (Chapter 4, Sec 4.3). Indeed, this algorithm converges to the first principal eigenvector of A.

Note that adaptive PCA algorithms are commonly used on streaming data and some GitHubs exist. I create here a comprehensive repository of these algorithms and also present new ones in this book and the associated GitHub.

For example, algorithm (1.4) has been widely used on streaming data to derive the strongest representation feature from the data for analytics and machine learning. Figure 1-6 shows multivariate non-stationary data with seasonality. Using the adaptive update rule (1.4), we derive the first principal component effectively in 0.2% of the streaming data. Figure 1-6 shows the seasonal time-varying streaming data on the left and the rapid convergence (ideal value is 1) of the algorithm on the right to the first principal eigenvector leading to the strongest data representation feature.

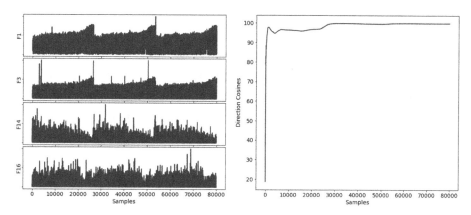

Figure 1-6. *Adaptive PCA algorithm used on seasonal streaming data*

The Python code to implement the Hebbian adaptive algorithm on real 33-dimensional dataset [nDim][nSamples] is

```
# Adaptive Hebbian/OJA algorithms
from numpy import linalg as la
A = np.zeros(shape=(nDim,nDim)) # stores adaptive
correlation matrix
```

```
w = 0.1 * np.ones(shape=(nDim)) # weight vectors of all
algorithms
for iter in range(nSamples):
    # update the data correlation matrix with latest data
    vector x
    x = np.array(dataset1.iloc[iter]).reshape(nDim,1)
    A = A + (1.0/(1 + iter))*((np.dot(x, x.T)) - A)
    # Hebbian/OJA Algorithm
    v = w[:].reshape(nDim, 1)
    v = v + (1/(100+iter))*(A @ v - v @ (v.T @ A @ v))
    w[:] = v.reshape(nDim)
```

Auto-Associative Networks

Auto-association is a neural network structure in which the desired output is same as the network input $\mathbf{x}_k$. This is also known as the linear autoencoder [autoencoder, Wikipedia]. Let's consider a two-layer linear network with weight matrices W_1 and W_2 for the input and output layers, respectively, and p ($\leq n$) nodes in the hidden layer. The mean square error (MSE) at the network output is given by

$$e = E\left[\| \mathbf{x}_k - W_2^T W_1^T \mathbf{x}_k \|^2\right]. \tag{1.5}$$

Due to p nodes in the hidden layer, a minimum MSE produces outputs that represent the best estimates of $\mathbf{x}_k \in \Re^n$ in the $\Re_p$ subspace. Since projection onto Φ_p minimizes the error in the $\Re_p$ subspace, we expect the first layer weight matrix W_1 to be rotations of Φ_p. The second layer weight matrix W_2 is the inverse of W_1 to finally represent the "best" identity transform. This intuitive argument has been proven [Baldi and Hornik 89,95, Bourland and Kamp 88], where the optimum weight matrices are

$$W_1 = \Phi_p R \text{ and } W_2 = R^{-1}\Phi_p^T, \tag{1.6}$$

where R is a non-singular $n \times n$ matrix. Note that if we further impose the constraint $W_2 = W_1^T$, then R is a unitary matrix and the input layer weight matrix W_1 is orthonormal and spans the space defined by Φ_p, the p principal eigenvectors of the input correlation matrix A. See Figure 1-7.

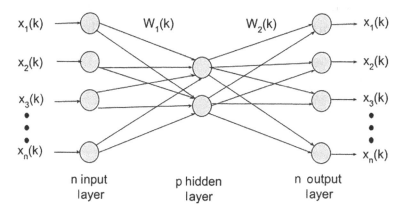

Figure 1-7. *Auto-associative neural network*

Note that if we have a single node in the hidden layer (i.e., $p=1$), then we obtain e as the output sum squared error for a two-layer linear auto-associative network with input layer weight vector $\mathbf{w}$ and output layer weight vector $\mathbf{w}^T$. The optimal value of $\mathbf{w}$ is the first principal eigenvector of the input correlation matrix A_k.

The result in (1.6) suggests the possibility of a PCA algorithm by using a gradient correction only to the input layer weights, while the output layer weights are modified in a symmetric fashion, thus avoiding the backpropagation of errors in one of the layers. One possible version of this idea is

$$W_1(k+1) = W_1(k) - \eta \frac{\partial e}{\partial W_1} \text{ and } W_2(k+1) = W_1(k+1)^T. \qquad (1.7)$$

Denoting $W_1(k)$ by W_k, we obtain an algorithm that is the same as Oja's *subspace learning algorithm* (SLA).

Algorithm (1.7) has been used widely on multivariate streaming data to extract the significant principal components so that we can instantaneously find the important data representation features. Figure 1-8 shows multidimensional streaming data on the left and the rapid convergence of the first two principal eigenvectors on the right by the adaptive algorithm (1.7), which is further described in Chapter 5.

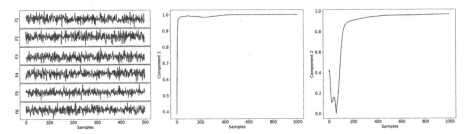

Figure 1-8. *Convergence of the first two principal eigenvectors computed from multidimensional streaming data. Data is on the left and feature convergence (ideal value = 1) is on the right*

The Python code to generate the first 4 principal eigenvectors from 10-dimensional synthetic dataset is

```python
from numpy import linalg as la
A = np.zeros(shape=(nDim,nDim)) # stores adaptive
correlation matrix
W2 = 0.1 * np.ones(shape=(nDim,nEA)) # weight vectors of all
algorithms
W3 = W2
c = [2-0.3*k for k in range(nEA)]
C = np.diag(c)
for epoch in range(nEpochs):
    for iter in range(nSamples):
        cnt = nSamples*epoch + iter
        # Update data correlation matrix A with current data
        sample x
```

```
x = X[:,iter]
x = x.reshape(nDim,1)
A = A + (1.0/(1 + cnt))*((np.dot(x, x.T)) - A)
# Deflated Gradient Descent
W2 = W2 + (1/(100 + cnt))*(A @ W2 - W2 @ np.triu(W2.T @
A @ W2))
# Weighted Gradient Descent
W3 = W3 + (1/(220 + cnt))*(A @ W3 @ C - W3 @ C @ (W3.T
@ A @ W3))
```

Hetero-Associative Networks

Let's consider a hetero-associative network, which differs from the auto-associative case in the output layer, which is **d** instead of **x**. Here **d** denotes the categorical classes the data belongs to. One example of $\mathbf{d}=\mathbf{e}_i$ where $\mathbf{e}_i$ is the ith standard basis vector [standard basis, Wikipedia] for class i. In a two-class problem, $\mathbf{d}=[1\ 0]^T$ for class 1 and $\mathbf{d}=[0\ 1]^T$ for class 2. Let's denote $B=E(\mathbf{xx}^T)$, $M=E(\mathbf{xd}^T)$, and $A=MM^T$. See Figure 1-9.

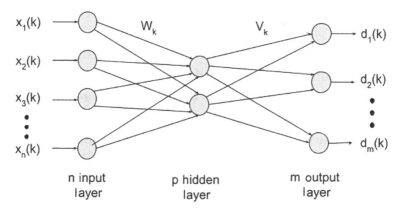

Figure 1-9. *Hetero-associative neural network*

We consider a two-layer linear hetero-associative neural network with just a single neuron in the hidden layer and $m \leq n$ output units. Let $\mathbf{w} \in \Re^n$

be the weight vector for the input layer and $\mathbf{v} \in \Re^m$ be the weight vector for the output layer. The MSE at the network output is

$$e = E\{\|\mathbf{d} - \mathbf{v}\mathbf{w}^T\mathbf{x}\|^2\}. \tag{1.8}$$

We further assume that the network has limited power, so $\mathbf{w}^T B \mathbf{w} = 1$. Hence, we impose this constraint on the MSE criterion in (1.8) as

$$J(\mathbf{w}, \mathbf{v}) = E\{\|\mathbf{d} - \mathbf{v}\mathbf{w}^T\mathbf{x}\|^2\} + \mu(\mathbf{w}^T B \mathbf{w} - 1), \tag{1.9}$$

where μ is the Lagrange multiplier. This equation has a unique global minimum where $\mathbf{w}$ is the first principal eigenvector of the matrix pencil (A, B) and $\mathbf{v} = M^T \mathbf{w}$. Furthermore, from the gradient of (1.9) with respect $\mathbf{w}$, we obtain the update equation for $\mathbf{w}$ as

$$\mathbf{w}_{k+1} = \mathbf{w}_k - \eta \frac{\partial J}{\partial \mathbf{w}}(\mathbf{w}_k, \mathbf{v}_k) = \mathbf{w}_k + \eta\left(I - B_k \mathbf{w}_k \mathbf{w}_k^T\right) M_k \mathbf{v}_k. \tag{1.10}$$

We can substitute the convergence value of $\mathbf{v}$ (1.10) and avoid the back-propagation of errors in the second layer to obtain

$$\mathbf{w}_{k+1} = \mathbf{w}_k + \eta\left(A_k \mathbf{w}_k - B_k \mathbf{w}_k\left(\mathbf{w}_k^T A_k \mathbf{w}_k\right)\right). \tag{1.11}$$

This algorithm can be used effectively to adaptively compute classification features from streaming data.

Figure 1-10 shows the multivariate e-shopping clickstream dataset [Apczynski M., et al.] belonging to two classes determining buyer's pricing sentiments. We use the adaptive algorithm (1.11) to compute the class separability feature $\mathbf{w}$. Figure 1-10 shows the following:

- The original multi-dimensional e-shopping data on the left 2 figures.

- The original data correlation on the right (3rd figure). The original data correlation matrix shows that the classes are indistinguishable in the original data.

- The correlation matrix of the data transformed by
  algorithm (1.11) on the far right. The white and red
  blocks on the far right matrix show the two classes
  are clearly separated in the transformed data by the
  algorithm (1.11).

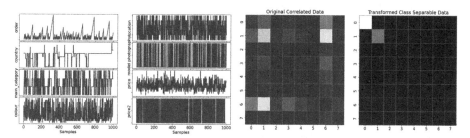

Figure 1-10. *e-Shopping clickstream data on the left and*
uncorrelated class separable data on the right

The Python code to generate the class separable transformed correlation
matrix from two-class multi-dimensional dataset[nDim, nSamples] is

```
# Adaptively compute matrices A and B
from numpy import linalg as la
nSamples = dataset1.shape[0]
nDim = dataset1.shape[1]
classes = np.array(dataset['price2']-1)
classes_categorical = tf.keras.utils.to_categorical(classes,
num_classes=2)
M = np.zeros(shape=(nDim,2)) # stores adaptive
correlation matrix
B = np.zeros(shape=(nDim,nDim)) # stores adaptive
correlation matrix
for iter in range(nSamples):
    cnt = iter + 1
    x = np.array(dataset1.iloc[iter])
```

```
    x = x.reshape(nDim,1)
    B = B + (1.0/cnt)*((np.dot(x, x.T)) - B)
    y = classes_categorical[iter].reshape(2,1)
    M = M + (1.0/cnt)*((np.dot(x, y.T)) - M)
    A = M @ M.T
# generate the transformed data
from scipy.linalg import eigh
eigvals, eigvecs = eigh(A, B)
V = np.fliplr(eigvecs)
VTAV = np.around(V.T @ A @ V, 2)
VTBV = np.around(V.T @ B @ V, 2)
# plot the LDA transformed data
import seaborn as sns
plt.figure(figsize=(12, 12))
plt.rcParams.update({'font.size': 16})
plt.subplot(2, 2, 1)
sns.heatmap(A, linewidth=0.5, linecolor="green", cmap='RdBu',
cbar=False)
plt.title("Original Correlated Data")
plt.subplot(2, 2, 2)
sns.heatmap(VTBV, linewidth=0.5, linecolor="green", cmap='hot',
cbar=False)
plt.title("Transformed Class Separable Data")
plt.subplot(2, 2, 3)
sns.heatmap(A, linewidth=0.5, linecolor="green", cmap='RdBu',
cbar=False)
plt.title("Original Correlated Data")
plt.subplot(2, 2, 4)
sns.heatmap(VTAV, linewidth=0.5, linecolor="green", cmap='hot',
cbar=False)
plt.title("Transformed Class Separable Data")
plt.show()
```

Statistical Pattern Recognition

One special case of linear hetero-association is a network performing one-from-m classification, where input $\mathbf{x}$ to the network is classified into one out of m classes $\omega_1,...,\omega_m$. If $\mathbf{x} \in \omega_i$ then $\mathbf{d}=\mathbf{e}_i$ where $\mathbf{e}_i$ is the ith standard basis vector. Unlike auto-associative learning, which is unsupervised, this network is supervised. In this case, A and B are scatter matrices. Here A is the *between-class scatter matrix* S_b, the scatter of the class means around the mixture mean, and B is the *mixture scatter matrix* S_m, the covariance of all samples regardless of class assignments. The generalized eigenvector decomposition of (A,B) is known as *linear discriminant analysis* (LDA), which was discussed in Section 1.1.

Information Theory

Another viewpoint of the data model (1.1) is due to Linsker [1988] and Plumbley [1993]. According to Linsker's *Infomax* principle, the optimum value of the weight matrix W is when the information $I(\mathbf{x},\mathbf{y})$ transmitted to its output $\mathbf{y}$ about its input $\mathbf{x}$ is *maximized*. This is equivalent to information in input $\mathbf{x}$ about output $\mathbf{y}$ since $I(\mathbf{x},\mathbf{y})=I(\mathbf{y},\mathbf{x})$. However, a noiseless process like (1.1) has infinite information about input $\mathbf{x}$ in $\mathbf{y}$ and vice versa since $\mathbf{y}$ perfectly represents $\mathbf{x}$. In order to proceed, we assume that input $\mathbf{x}$ contains some noise $\mathbf{n}$, which prevents $\mathbf{x}$ from being measured accurately by $\mathbf{y}$. There are two variations of this model, both inspired by Plumbley [1993].

In the first model, we assume that the output $\mathbf{y}$ is corrupted by noise due to the transform W. We further assume that average power available for transmission is limited, the input $\mathbf{x}$ is zero-mean Gaussian with covariance A, the noise $\mathbf{n}$ is zero-mean uncorrelated Gaussian with covariance N, and the transform noise $\mathbf{n}$ is independent of $\mathbf{x}$. The noisy data model is

$$\mathbf{y} = W^T\mathbf{x} + \mathbf{n}. \tag{1.12}$$

The mutual information $I(\mathbf{y}, \mathbf{x})$ (with Gaussian assumptions) is

$$I(\mathbf{y}, \mathbf{x}) = 0.5 \log \det(W^T A W + N) - 0.5 \log \det(N). \qquad (1.13)$$

We define an objective function with power constraints as

$$J(W) = I(\mathbf{y}, \mathbf{x}) - 0.5tr(\Lambda(W^T A W + N)), \qquad (1.14)$$

where Λ is a diagonal matrix of Lagrange multipliers. This function is maximized when $W = \Phi R$, where Φ is the principal eigenvector matrix of A and R is a non-singular rotation matrix.

Optimization Theory

In optimization theory, various matrix functions are computed by evaluating the maximums and minimums of objective functions. Given a symmetric positive definite matrix pencil (A, B), the first principal generalized eigenvector is obtained by maximizing the well-known Raleigh quotient:

$$J(\mathbf{w}) = \frac{\mathbf{w}^T A \mathbf{w}}{\mathbf{w}^T B \mathbf{w}}. \qquad (1.15)$$

There are three common modifications of this objective function based on the method of optimization [Luenberger]. They are

- Lagrange multiplier:

$$J(\mathbf{w}) = -\mathbf{w}^T A \mathbf{w} + \alpha(\mathbf{w}^T B \mathbf{w} - 1), \qquad (1.16)$$

where α is a Lagrange multiplier.

- Penalty function:

$$J(\mathbf{w}) = -\mathbf{w}^T A \mathbf{w} + \mu(\mathbf{w}^T B \mathbf{w} - 1)^2, \qquad (1.17)$$

where μ is a non-negative scalar constant.

- Augmented Lagrangian:

$$J(\mathbf{w}) = -\mathbf{w}^T A \mathbf{w} + \alpha(\mathbf{w}^T B \mathbf{w} - 1) + \mu(\mathbf{w}^T B \mathbf{w} - 1)^2, \qquad (1.18)$$

  where α is a Lagrange multiplier and $\mu > 0$ is the penalty constant. Several algorithms based on this objective function are given in Chapters 4-7.

We obtain adaptive algorithms from these objective functions by using instantaneous values of the gradients and a gradient ascent technique, as discussed in the following chapters. One advantage of these objective functions is that we can use accelerated convergence methods such as steepest descent, conjugate direction, and Newton-Raphson. Chapter 6 discusses several such methods. The recursive least squares method has also been applied to adaptive matrix computations by taking various approximations of the inverse Hessian of the objective functions.

1.3 Why Adaptive Algorithms?

We observed that data representation and classification problems lead to matrix algebra problems, which have two solutions depending on the nature of the inputs.

1. The first set of solutions requires the relevant matrices to be known in advance.

2. The second set of solutions requires a sequence of samples from which the matrix can be computed.

In this section, I explain the difference between batch and adaptive processing, benefits of adaptive processing on streaming data, and the requirements for adaptive algorithms.

Iterative or Batch Processing of Static Data

When the data is available in advance and the underlying matrices are known, we can use the batch processing approach. These algorithms have been used to solve a large variety of matrix algebra problems such as matrix inversion, EVD, SVD, and PCA [Cichocki et al. 92, 93]. These algorithms are solved in two steps: (1) using the pooled data to estimate the required matrices, and (2) using a numerical matrix algebra procedure to solve the necessary matrix functions.

For example, consider a simple matrix inversion problem for the correlation matrix A of a data sequence $\{\mathbf{x}_k\}$. We can calculate the matrix inverse A^{-1} after all of the data have been collected and A has been calculated. This approach works in a batch fashion. When a new sample $\mathbf{x}$ is added, it is not difficult to get the inverse of the new matrix $A_{new}=(nA+\mathbf{x}\mathbf{x}^T)/(n+1)$, where n is the total number of samples used to compute A. Although the computation for A is simple, all the computations for solving A_{new}^{-1} need to be repeated.

There are two problems with this batch processing approach.

1. First, the dimension of the samples may be large so that even if all the samples are available, performing the matrix algebra may be difficult or may take a prohibitively large amount of computational time. For example, eigenvector evaluation requires $O(n^3)$ computation, which is infeasible for samples of a large dimension (say 1,000) which occurs commonly in image processing and automatic control applications.

2. Second, the matrix functions evaluated by conventional schemes cannot adapt to small changes in the data (e.g., a few incoming samples). If the matrix functions are estimated by conventional methods from K samples, then for each additional sample all of the computation has to be repeated.

These deficiencies make the batch schemes inefficient for real-time applications where the data arrives incrementally or in an *online* fashion.

My Approach: Adaptive Processing of Streaming Data

When we have a continuous sequence of data samples, batch algorithms are no longer useful. Instead, adaptive algorithms are used to solve matrix algebra problems. An immediate advantage of these algorithms is that they can be used on real-time problems such as edge computation, adaptive data compression [Le Gall 91], antenna array processing for noise analysis and source location [Owsley 78], and adaptive spectral analysis for frequency estimation [Pisarenko 73].

My approach is to offer computationally simple adaptive algorithms for matrix computations from streaming data. Note that adaptive algorithms are critical in environments where the data volume is large, the data has high dimensions, the data is time varying and has changing underlying statistics, and we do not have sufficient storage, compute, and bandwidth to process the data with low latency. One such environment is edge devices and computation.

For example, given streaming samples $\{\mathbf{x}_k\}$ of customer e-shopping data, we need to calculate a key customer sentiment $\mathbf{w}_k$ from the data stream. For that, we need to design a **simple update rule** to change the customer sentiment $\mathbf{w}$ as new data $\mathbf{x}$ is available. The simple update rule will change the sentiment $\mathbf{w}_k$ to its latest value $\mathbf{w}_{k+1}$ as new data $\mathbf{x}_k$ is available. It is of the format $\mathbf{w}_{k+1} = \mathbf{w}_k + f(\mathbf{w}_k, \mathbf{x}_k)$, where

- $\mathbf{w}_k$ is last value of the sentiment,
- $\mathbf{w}_{k+1}$ is the latest updated value of the sentiment,
- $\mathbf{x}_k$ is the newest data used to calculate $\mathbf{w}_{k+1}$, and
- $f(.)$ is a simple function of $\mathbf{w}_k$ and $\mathbf{x}_k$.

An example of this update rule is the well-known algorithm [Oja 82] to compute the first principal eigenvector of a streaming sequence {$\mathbf{x}_k$}:

$$\mathbf{w}_{k+1} = \mathbf{w}_k + \eta \left(\mathbf{x}_k \mathbf{x}_k^T - \mathbf{w}_k \mathbf{w}_k^T \mathbf{x}_k \mathbf{x}_k^T \right) \mathbf{w}_k , \qquad (1.19)$$

where η>0 is a small gain constant. In this algorithm, for each sample $\mathbf{x}_k$ the update procedure requires simple matrix-vector multiplications, and the vector $\mathbf{w}_k$ converges to the principal eigenvector of the data correlation matrix A. Clearly, this can be easily implemented in small CPUs.

Figure 1-11 shows a multivariate e-shopping clickstream dataset [Apczynski M. et al.]. The adaptive update rule (1.19) is used to compute buyer pricing sentiments. The data is shown on the left and sentiments computed adaptively are shown on the right (ideal value is 1). The sentiments are updated adaptively as new data arrives and the sentiment value converges quickly to its ideal value of 1.

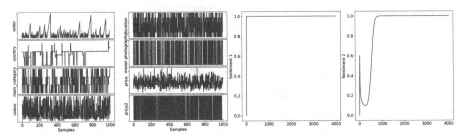

Figure 1-11. *e-Shopping clickstream data on the left and buyer sentiments computed by the update rule (1.19) on the right (ideal value = 1)*

There are several advantages and disadvantages of adaptive algorithms over conventional iterative batch solutions.

The advantages are

- While conventional solutions find all eigenvectors and eigenvalues, adaptive solutions compute the desired eigenvectors only. In many applications, we do not need eigenvalues. Hence, they should be more efficient.

- Furthermore, computational complexity of adaptive algorithms depends on the efficiency of the matrix-vector product $A\mathbf{x}$. These methods can be even more efficient when the matrix A has a certain structure such as sparse, Hankel, or Toeplitz for which FFT can be used to speed up the computation.

- Adaptive algorithms easily fit the framework of time-varying multidimensional processes such as adaptive signal processing, where the input process is continuously updated.

The disadvantages are

- Adaptive algorithms produce an approximate value of the features for the current dataset whereas batch algorithms provide an exact value.

- The adaptive approach requires a "ramp up" time to reach high accuracy as evidenced by the curves in Figure 1-10.

Requirements of Adaptive Algorithms

It is clear from the previous discussion that adaptive algorithms for matrix computation need to be inexpensive for the computational process to keep pace with the input data stream. We also require that the estimates converge strongly to their actual values. We therefore expect our adaptive algorithms to satisfy the following constraints:

- The algorithms should adapt to small changes in data, which is useful for real-time applications.

- The estimates obtained from the algorithms should have known statistical properties.

- The network architectures associated with the adaptive algorithms consist of linear units in one- or two-layer networks, such that the networks can be easily implemented with simple CPUs and no special hardware.

- The computation involved in the algorithms is inexpensive such that the statistical procedure can process every data sample as they arrive.

- The estimates obtained from the algorithms converge strongly to their actual values.

With these requirements, we proceed to design adaptive algorithms that solve the matrix algebra problems considered in this book. The objective of this book is to develop a variety of neuromorphic adaptive algorithms to solve matrix algebra problems.

Although I will discuss many adaptive algorithms to solve the matrix functions, most of the algorithms are of the following general form:

$$W_{k+1} = W_k + \eta H(\mathbf{x}_k, W_k), \tag{1.20}$$

where $H(\mathbf{x}_k, W_k)$ follows certain continuity and regularity properties [Ljung 77,78,84,92], and $\eta > 0$ is a small gain constant. These algorithms satisfy the requirements outlined before.

Real-World Use of Adaptive Matrix Computation Algorithms and GitHub

In Chapter 8, I discuss several real-world applications of these adaptive matrix computation algorithms. I also published the code for these applications in a public GitHub [Chanchal Chatterjee GitHub].

1.4 Common Methodology for Derivations of Algorithms

My contributions in this book are two-fold:

1. I present a common methodology to derive and analyze each adaptive algorithm.

2. I present adaptive algorithms to a number of matrix algebra problems.

The literature for adaptive algorithms for matrix computation offers a wide range of techniques (including ad hoc methods) and various types of convergence procedures. In this book, I present a *common methodology* to derive and prove the convergence of the adaptive algorithms (all proofs are in the GitHub).

The advantage of adopting this is to allow the reader to follow the methodology and derive new adaptive algorithms for their use cases.

In the following chapters, I follow the following steps to derive each algorithm:

1. **Objective function**

   I first present an *objective function* $J(W;A_k)$ such that the minimizer W^* of J is the desired matrix function of the data matrix A.

2. **Derive the adaptive algorithm**

   I derive an *adaptive update rule* for matrix W by applying the gradient descent technique on the objective function $J(W;A_k)$. The adaptive gradient descent update rule is

$$W_{k+1} = W_k - \eta_k \nabla_W J(W_k, A_k) = W_k + \eta_k h(W_k, A_k), \tag{1.21}$$

29

where the function $h(W_k, A_k)$ follows certain continuity and regularity properties and η_k is a decreasing gain sequence.

3. **Speed up the adaptive algorithm**

   The availability of the objective function $J(W;A)$ allows us to speed up the adaptive algorithm by applying speedup techniques in optimization theory such as steepest descent, conjugate direction, Newton-Raphson, and recursive least squares. Details of these methods for principal component analysis are given in Chapter 6.

4. **Show the algorithms converge to the matrix functions**

   In this book, I provide numerical experiments to demonstrate the convergence of these algorithms. The mathematical proofs are provided in a separate document in the GitHub [Chatterjee Github].

An important benefit of following this methodology is to allow practitioners to derive new algorithms for their own use cases.

Matrix Algebra Problems Solved Here

In the applications, I consider the data as arriving in temporal succession, such as in vector sequences $\{\mathbf{x}_k\}$ or $\{\mathbf{y}_k\}$, or in matrix sequences $\{A_k\}$ or $\{B_k\}$. Note that the algorithms given here can be used for non-stationary input streams.

In the following chapters, I present novel adaptive algorithms to estimate the following matrix functions:

- Mean, correlation, and normalized mean of $\{\mathbf{x}_k\}$

- Square root ($A^{\frac{1}{2}}$) and inverse of the square root ($A^{-\frac{1}{2}}$) of A from $\{\mathbf{x}_k\}$ or $\{A_k\}$

- Principal eigenvector of A from $\{\mathbf{x}_k\}$ or $\{A_k\}$

- Principal and minor eigenvector of A from $\{\mathbf{x}_k\}$ or $\{A_k\}$

- Generalized eigenvectors of A with respect to B from $\{\mathbf{x}_k\}$ and $\{\mathbf{z}_k\}$ or $\{A_k\}$ and $\{B_k\}$

- Singular value decomposition (SVD) of C from $\{\mathbf{x}_k\}$ and $\{\mathbf{z}_k\}$ or $\{C_k\}$

Besides these algorithms, I also discuss

- Adaptive computation of mean, covariance, and matrix inversion

- Methods to accelerate the adaptive algorithms by techniques of nonlinear optimization

1.5 Outline of The Book

In the following chapters, I discuss in detail the formulation, derivation, convergence, and experimental results of many adaptive algorithms for various matrix algebra problems.

In Chapter 2, I discuss basic terminologies and methods used throughout the remaining chapters. This chapter also discusses adaptive algorithms mean, median, correlation, covariance, and inverse correlation/covariance computation for both stationary and non-stationary data. I further discusses novel adaptive algorithms for normalized mean computation.

In Chapter 3, I discuss three adaptive algorithms for the computation of the square root of a matrix sequence. I next discuss three algorithms for the inverse square root of the same. I offer objective functions and convergence proofs for these algorithms.

In Chapter 4, I discuss 11 algorithms, some of them new algorithms, for the adaptive computation of the first principal eigenvector of a matrix sequence or the online correlation matrix of a vector sequence. I offer best practices to choose the algorithm for a given application. I follow the common methodology of deriving and analyzing the convergence of each algorithm, supported by experimental results.

In Chapter 5, I present 21 adaptive algorithms for the computation of principal and minor eigenvectors of a matrix sequence or the online correlation matrix of a vector sequence. These algorithms are derived from 7 different types of objective functions, each under 3 different conditions. Each algorithm is derived, discussed, and shown to converge analytically and experimentally. I offer best practices to choose the algorithm for a given application.

Since I have objective functions for all of the adaptive algorithms, in Chapter 6, I deviate from the traditional gradient descent method of deriving the algorithms. Here I derive new computationally faster algorithms by using steepest descent, conjugate direction, Newton-Raphson, and recursive least squares on the objective functions. Experimental results and comparison with state-of-the-art algorithms show the faster convergence of these adaptive algorithms.

In Chapter 7, I discuss 21 adaptive algorithms for generalized eigen-decomposition from two matrix or vector sequences. Once again, I follow the common methodology and derive all algorithms from objective functions, followed by experimental results.

In Chapter 8, I present real-world applications of these algorithms with examples and code.

The bibliography is in Chapter 9.

CHAPTER 2

General Theories and Notations

2.1 Introduction

In this chapter, I present algorithms for the adaptive solutions of matrix algebra problems from a sequence of matrices. The streams or sequences can be random matrices $\{A_k\}$ or $\{B_k\}$, or the correlation matrices of random vector sequences $\{\mathbf{x}_k\}$ or $\{\mathbf{y}_k\}$. Examples of matrix algebra are matrix inversion, square root, inverse square root, eigenvectors, generalized eigenvectors, singular vectors, and generalized singular vectors.

This chapter additionally covers the basic terminologies and methods used throughout the remaining chapters. I also present well-known adaptive algorithms to compute the mean, median, covariance, inverse covariance, and correlation matrices from random matrix or vector sequences. Furthermore, I present a new algorithm to compute the normalized mean of a random vector sequence.

For the sake of simplicity, let's assume that the multidimensional data $\{\mathbf{x}_k \in \Re^n\}$ arrives as a sequence. From this data sequence, we can derive a matrix sequence $\{A_k = \mathbf{x}_k \mathbf{x}_k^T\}$. We define the data correlation matrix A as follows:

$$A = \lim_{k \to \infty} E\left[\mathbf{x}_k \mathbf{x}_k^T\right]. \tag{2.1}$$

© Chanchal Chatterjee 2022
C. Chatterjee, *Adaptive Machine Learning Algorithms with Python*,
https://doi.org/10.1007/978-1-4842-8017-1_2

2.2 Stationary and Non-Stationary Sequences

In practical implementations, we face two types of sequences: stationary and non-stationary. A sequence $\{\mathbf{x}_k\}$ is considered *asymptotically* (*weak*) *stationary* if $\lim_{k\to\infty} E\left[\mathbf{x}_k\mathbf{x}_k^T\right]$ is a constant. For a *non-stationary* sequence $\{\mathbf{x}_k\}$, $E\left[\mathbf{x}_k\mathbf{x}_{k+m}^T\right]$ remains a function of both k and m, and $E\left[\mathbf{x}_k\mathbf{x}_k^T\right]$ is a function of k. Examples of non-stationary data are given in *Publicly Real-World Datasets to Evaluate Stream Learning Algorithms* [Vinicius Souza et al. 20]. Figure 2-1 shows examples of stationary and non-stationary data.

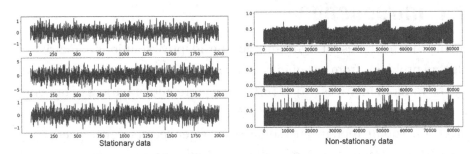

Figure 2-1. *Examples of stationary and non-stationary data*

Non-stationarity in data can be detected by well-known techniques described in this reference [Shay Palachy 19].

2.3 Use Cases for Adaptive Mean, Median, and Covariances

Adaptive mean computation is important in real-world applications even though it is one of the simplest algorithms.

Handwritten Character Recognition

Consider the problem of detecting handwritten number 8. Figure 2-2 shows six instances of the number 8 from the Keras dataset MNIST images [Keras, MNIST].

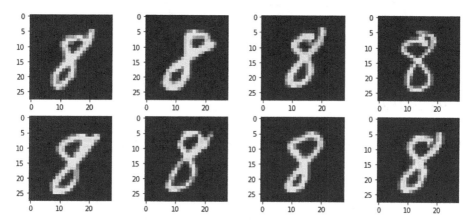

Figure 2-2. *Examples of variations in handwritten number 8*

One algorithm to recognize the number 8 with all its variations is to find the mean set of pixels that represent it. Due to random variations in handwriting, it is difficult to compile all variations of each character ahead of time. It is important to design an adaptive learning scheme whereby instantaneous variations in these characters are immediately represented in the machine learning algorithm.

The adaptive mean detection algorithm is helpful in finding the common pattens that represent the number 8. Figure 2-3 shows a template for the number 8 obtained by the adaptive mean algorithm given in Eq. (2.2).

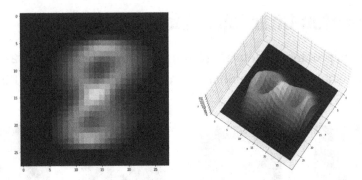

Figure 2-3. *Template for number 8. 2D representation on the left and 3D on the right*

Anomaly Detection of Streaming Data

A simple yet powerful algorithm to detect anomalies in data is to calculate the median and compare the current value against the median of the data. Figure 2-4 shows streaming data [Yahoo Research Webscope S5 Data] containing occasional anomalous values. We adaptively computed the median with algorithm (2.20) and compared it against the data to detect anomalies. Here the data samples are in blue, the adaptive median is in green, and anomalies are in red.

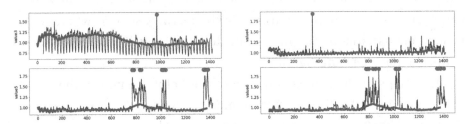

Figure 2-4. *Anomalies detected with an adaptive median algorithm on time series data*

More on this topic is discussed in Chapter 8.

2.4 Adaptive Mean and Covariance of Nonstationary Sequences

In the stationary case, given a sequence $\{\mathbf{x}_k \in \Re^n\}$, we can compute the adaptive mean $\mathbf{m}_k$ as follows:

$$\mathbf{m}_k = \frac{1}{k}\sum_{i=1}^{k}\mathbf{x}_i = \mathbf{m}_{k-1} + \frac{1}{k}\left(\mathbf{x}_k - \mathbf{m}_{k-1}\right). \tag{2.2}$$

Similarly, the adaptive correlation A_k is

$$A_k = \frac{1}{k}\sum_{i=1}^{k}\mathbf{x}_i\mathbf{x}_i^T = A_{k-1} + \frac{1}{k}\left(\mathbf{x}_k\mathbf{x}_k^T - A_{k-1}\right). \tag{2.3}$$

Here we use all samples up to time instant k.

If, however, the data is non-stationary, we use a *forgetting factor* $0 < \beta \le 1$ to implement an effective window of size $1/(1-\beta)$ as

$$\mathbf{m}_k = \frac{1}{k}\sum_{i=1}^{k}\beta^{k-i}\mathbf{x}_i = \beta\mathbf{m}_{k-1} + \frac{1}{k}\left(\mathbf{x}_k - \beta\mathbf{m}_{k-1}\right) \tag{2.4}$$

and

$$A_k = \frac{1}{k}\sum_{i=1}^{k}\beta^{k-i}\mathbf{x}_i\mathbf{x}_i^T = \beta A_{k-1} + \frac{1}{k}\left(\mathbf{x}_k\mathbf{x}_k^T - \beta A_{k-1}\right). \tag{2.5}$$

This effective window ensures that the past data samples are downweighted with an exponentially fading window compared to the recent ones in order to afford the tracking capability of the adaptive algorithm. The exact value of β depends on the specific application. Generally speaking, for slow time-varying $\{\mathbf{x}_k\}$, β is chosen close to 1 to implement a large effective window, whereas for fast time-varying $\{\mathbf{x}_k\}$, β is chosen near zero for a small effective window [Benveniste et al. 90].

The following is Python code to adaptively compute a mean vector and correlation matrix with data X[nDim,nSamples]:

```
for epoch in range(nEpochs):
    for iter in range(nSamples):
        cnt = nSamples*epoch + iter
        x = X[:,iter]
        x = x.reshape(nDim,1)
        # Eq.2.4
        m = beta * m  + (1.0/(1 + cnt)) * (x - beta * m)
        # Eq.2.5
        A = beta * A + (1.0/(1 + cnt))*((np.dot(x, x.T)) -
        beta * A)
```

2.5 Adaptive Covariance and Inverses

Given the mean and correlations discussed before, the adaptive covariance matrix B_k can also be computed as follows:

$$B_k = \frac{1}{k}\sum_{i=1}^{k}\beta^{k-i}\left(\mathbf{x}_i - \mathbf{m}_i\right)\left(\mathbf{x}_i - \mathbf{m}_i\right)^T = \beta B_{k-1}$$
$$+ \frac{1}{k}\left(\left(\mathbf{x}_k - \mathbf{m}_k\right)\left(\mathbf{x}_k - \mathbf{m}_k\right)^T - \beta B_{k-1}\right). \tag{2.6}$$

From the adaptive correlation matrix A_k in (2.5), the inverse correlation matrix A_k^{-1} can be obtained adaptively by the Sherman-Morrison formula [Sherman–Morrison, Wikipedia] as

$$A_k^{-1} = \frac{k}{\beta(k-1)}\left(A_{k-1}^{-1} - \frac{A_{k-1}^{-1}\mathbf{x}_k\mathbf{x}_k^T A_{k-1}^{-1}}{\beta(k-1)+\mathbf{x}_k^T A_{k-1}^{-1}\mathbf{x}_k}\right). \tag{2.7}$$

Similarly, the inverse covariance matrix B_k^{-1} can be obtained adaptively as

$$B_k^{-1} = \frac{k}{\beta(k-1)}\left(B_{k-1}^{-1} - \frac{B_{k-1}^{-1}(\mathbf{x}_k - \mathbf{m}_k)(\mathbf{x}_k - \mathbf{m}_k)^T B_{k-1}^{-1}}{\beta(k-1) + (\mathbf{x}_k - \mathbf{m}_k)^T B_{k-1}^{-1}(\mathbf{x}_k - \mathbf{m}_k)}\right). \qquad (2.8)$$

The following is Python code to adaptively compute inverse correlation and inverse covariance matrices with data X[nDim,nSamples]:

```
for epoch in range(nEpochs):
    for iter in range(nSamples):
        cnt = nSamples*epoch + iter
        x = X[:,iter]
        x = x.reshape(nDim,1)
        y = Y[:,iter]
        y = x.reshape(nDim,1)
        # Eq.2.7
        k = cnt+2
        AW = (k/(beta*(k-1))) * (AW - (AW * (x @ x.T) * AW) \
                            / (beta*(k-1) + x.T @ AW @ x))
        # Eq.2.8
        BW = (k/(beta*(k-1))) * (BW - (BW * (y @ y.T) * BW) \
                            / (beta*(k-1) + y.T @ BW @ y))
```

2.6 Adaptive Normalized Mean Algorithm

The most obvious choice for an adaptive normalized mean algorithm is to use (2.2) and normalize each $\mathbf{m}_k$. However, a more efficient algorithm can be obtained from the following cost function whose minimizer $\mathbf{w}^*$ is the asymptotic normalized mean $\mathbf{m}/\|\mathbf{m}\|$, where $\mathbf{m} = \lim_{k \to \infty} E[\mathbf{x}_k]$:

$$J(\mathbf{w}_k;\mathbf{x}_k) = \|\mathbf{x}_k - \mathbf{w}_k\|^2 + \alpha(\mathbf{w}_k^T \mathbf{w}_k - 1), \qquad (2.9)$$

where α is a Lagrange multiplier that enforces the constraint that the mean is normalized. The gradient of $J(\mathbf{w}_k;\mathbf{x}_k)$ with respect to $\mathbf{w}_k$ is

$$(1/2)\nabla_{\mathbf{w}_k}J\left(\mathbf{w}_k;\mathbf{x}_k\right)=-\left(\mathbf{x}_k-\mathbf{w}_k\right)+\alpha\mathbf{w}_k. \tag{2.10}$$

Multiplying (2.10) by $\mathbf{w}_k^T$ and applying the constraint $\mathbf{w}_k^T\mathbf{w}_k=1$, we obtain

$$\alpha=\mathbf{w}_k^T\mathbf{x}_k-1. \tag{2.11}$$

Using this α in (2.11), we obtain the adaptive gradient descent algorithm for normalized mean:

$$\mathbf{w}_{k+1}=\mathbf{w}_k+\eta_k\left(\mathbf{x}_k-\mathbf{w}_k^T\mathbf{x}_k\mathbf{w}_k\right), \tag{2.12}$$

where η_k is a small decreasing constant, which follows assumption A1.2 in the Proofs of Convergence in GitHub [Chatterjee GitHub].

Variations of the Adaptive Normalized Mean Algorithm

There are several variations of the objective function (2.9) that lead to many other adaptive algorithms for normalized mean computation. One variation of (2.9) is to place the value of α in (2.11) in the objective function (2.9) to obtain the following objective function:

$$J\left(\mathbf{w}_k;\mathbf{x}_k\right)=\left\|\mathbf{x}_k-\mathbf{w}_k\right\|^2+\left(\mathbf{w}_k^T\mathbf{x}_k-1\right)\left(\mathbf{w}_k^T\mathbf{w}_k-1\right). \tag{2.13}$$

Unlike (2.9), this objective function is unconstrained and has the constraint $\mathbf{w}_k^T\mathbf{w}_k=1$ built into it. It leads to the following adaptive algorithm:

$$\mathbf{w}_{k+1}=\mathbf{w}_k+\eta_k\left(2\mathbf{x}_k-\mathbf{w}_k^T\mathbf{x}_k\mathbf{w}_k-\mathbf{w}_k^T\mathbf{w}_k\mathbf{x}_k\right). \tag{2.14}$$

Another variation is the use of a penalty function method of nonlinear optimization that enforces the constraint $\mathbf{w}_k^T \mathbf{w}_k = 1$. This objective function is

$$J\left(\mathbf{w}_k; \mathbf{x}_k\right) = \left\|\mathbf{x}_k - \mathbf{w}_k\right\|^2 + \frac{\mu}{2}\left(\mathbf{w}_k^T \mathbf{w}_k - 1\right)^2, \tag{2.15}$$

where μ is a positive penalty constant. This objective function is also unconstrained and leads to the following adaptive algorithm:

$$\mathbf{w}_{k+1} = \mathbf{w}_k + \eta_k\left(\mathbf{x}_k - \mathbf{w}_k - \mu \mathbf{w}_k\left(\mathbf{w}_k^T \mathbf{w}_k - 1\right)\right). \tag{2.16}$$

The following is the Python code to adaptively compute a normalized mean by algorithms (2.12), (2.14), and (2.16) with data X[nDim,nSamples]:

```
for epoch in range(nEpochs):
    for iter in range(nSamples):
        cnt = nSamples*epoch + iter
        x = X[:,iter]
        x = x.reshape(nDim,1)
        # Eq.2.12
        w1 = w1 + (1.0/(100+cnt))*(x - (w1.T @ x)*w1)
        # Eq.2.14
        w2 = w2 + (1.0/(100+cnt))*(2*x-(w2.T @ x)*w2 - (w2.T
        @ w2)*x)
        # Eq.2.16
        w3 = w3 + (1.0/(100+cnt))*(x - w3- mu* w3 @ ((w3.T
        @ w3)-1))
```

2.7 Adaptive Median Algorithm

Given a sequence {$\mathbf{x}_k$}, its asymptotic median μ satisfies the following:

$$\lim_{k \to \infty} P\left(\mathbf{x}_k \geq \mu\right) = \lim_{k \to \infty} P\left(\mathbf{x}_k < \mu\right) = 0.5, \tag{2.17}$$

where P(E) is the probability measure of event E and $0 \leq P(E) \leq 1$.
The objective function $J(\mathbf{w}_k;\mathbf{x}_k)$ whose minimizer $\mathbf{w}^*$ is the asymptotic
median μ is

$$J(\mathbf{w}_k;\mathbf{x}_k) = \|\mathbf{x}_k - \mathbf{w}_k\|. \tag{2.18}$$

The gradient of $J(\mathbf{w}_k;\mathbf{x}_k)$ with respect to $\mathbf{w}_k$ is

$$\nabla_{\mathbf{w}_k} J\left(\mathbf{w}_k;\mathbf{x}_k\right) = -\mathrm{sgn}\left(\mathbf{x}_k - \mathbf{w}_k\right), \tag{2.19}$$

where sgn(.) is the sign operator (sgn(x)=1 if $x \geq 0$ and –1 if $x < 0$). From
the gradient in (2.19), we obtain the adaptive gradient descent algorithm:

$$\mathbf{w}_{k+1} = \mathbf{w}_k + \eta_k \, \mathrm{sgn}\,(\mathbf{x}_k - \mathbf{w}_k). \tag{2.20}$$

The following is the Python code to adaptively compute the median by
algorithm (2.20) with data X[nDim,nSamples]:

```
for epoch in range(nEpochs):
    for iter in range(nSamples):
        cnt = nSamples*epoch + iter
        x = X[:,iter]
        x = x.reshape(nDim,1)
        # Eq.2.20
        md = md + (3.0/(1 + cnt)) * np.sign(x - md)
```

2.8 Experimental Results

The purpose of these experiments is to demonstrate the performance
and accuracies of the adaptive algorithms for mean, correlation, inverse
correlation, inverse covariance, normalized mean, and median.

I generated 1,000 sample vectors $\{\mathbf{x}_k\}$ of five-dimensional Gaussian data (i.e., $n=5$) with the following mean and covariance:

$$\text{Mean} = [10\ 7\ 6\ 5\ 1],$$

$$\text{Covariance} = \begin{bmatrix} 2.091 & 0.038 & -0.053 & -0.005 & 0.010 \\ 0.038 & 1.373 & 0.018 & -0.028 & -0.011 \\ -0.053 & 0.018 & 1.430 & 0.017 & 0.055 \\ -0.005 & -0.028 & 0.017 & 1.084 & -0.005 \\ 0.010 & -0.011 & 0.055 & -0.005 & 1.071 \end{bmatrix}.$$

For each algorithm, I computed the error as the Frobenius norm [Frobenius norm, Wikipedia] of the estimated value at each iteration of the algorithm and the actual computed value from all of the 1,000 samples.

$$error(k) = \left\| Estimated\ Value(k) - Actual\ Value \right\|_F .$$

I computed the mean vector and correlation matrices by the adaptive algorithms (2.4) and (2.5), respectively. Figure 2-5 shows the results.

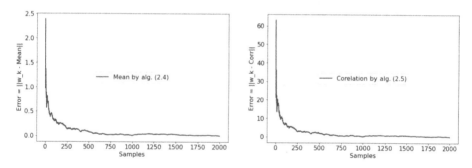

Figure 2-5. *Convergence of the mean vector and correlation matrices with algorithms (2.4) and (2.5), respectively*

I computed the inverses of the correlation and covariance matrices with algorithms (2.7) and (2.8), respectively. Figure 2-6 shows the results.

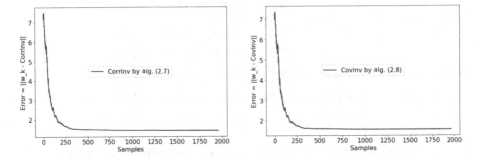

Figure 2-6. *Convergence of the inverse correlation and inverse covariance matrices with algorithms (2.7) and (2.8), respectively*

I computed the normalized mean by adaptive algorithms (2.12) and (2.14), and the median by algorithm (2.20). For each value of k in these algorithms, I computed the errors between the $\mathbf{w}_k$ (estimate) and the actual values of normalized mean and median obtained from the entire 1,000 sample data. I used the 5X1 zero vector as the starting values ($\mathbf{w}_0$) for all algorithms. The results are shown in Figure 2-7.

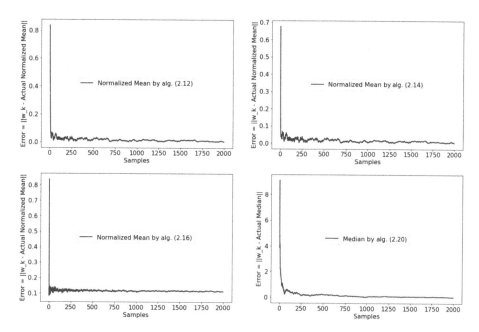

Figure 2-7. *Convergence of the normalized mean of {$\mathbf{x}_k$} by adaptive algorithms (2.12), (2.14), and (2.16) and the median by adaptive algorithm (2.20)*

All adaptive algorithms converged rapidly. After the 1,000 samples were processed, the errors were 0.0043 for algorithms (2.14) and (2.16), and 0.0408 for algorithm (2.20). See the code in the GitHub repository [Chatterjee GitHub].

CHAPTER 3

Square Root and Inverse Square Root

3.1 Introduction and Use Cases

Adaptive computation of the square root and inverse square root of the real-time correlation matrix of a streaming sequence $\{\mathbf{x}_k \in \Re^n\}$ has numerous applications in machine learning, data analysis, and image/signal processing. They include data whitening, classifier design, and data normalization [Foley and Sammon 75; Fukunaga 90].

Data whitening is a process of decorrelating the data such that all components have unit variance. It is a data preprocessing step in machine learning and data analysis to "normalize" the data so that it is easier to model. Prominent applications are

- To transform correlated noise in a signal to independent and identically distributed (iid) noise, which is easier to classify

- Generalized eigenvector computation [Chatterjee et al. Mar 97] (see Chapter 7)

C. Chatterjee, *Adaptive Machine Learning Algorithms with Python*, https://doi.org/10.1007/978-1-4842-8017-1_3

- Linear discriminant analysis computation [linear discriminant analysis, Wikipedia]

- Gaussian classifier design and computation of distance measures [Chatterjee et al. May 97]

Figure 3-1 shows the correlation matrices of the original and whitened data. The original data is highly correlated as shown by the colors on all axes. The whitened data is fully uncorrelated with no correlation between components since only diagonal values exist.

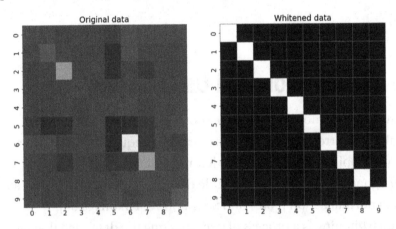

Figure 3-1. *Original correlated data on the left and the uncorrelated "whitened" data on the right*

Figure 3-2 shows a handwritten number 0 obtained from the Keras MNIST dataset [Keras, MNIST]. The correlation matrix on the right shows that the data pixels are highly correlated for all pixels.

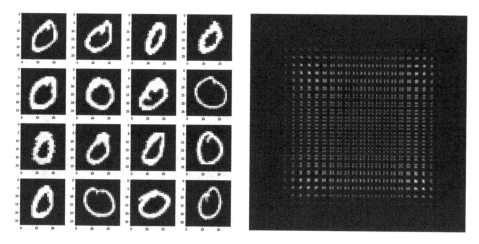

Figure 3-2. *Handwritten MNIST number 0 and the correlation matrix of all characters*

The following Python code whitens the data samples X[nDim,nSamples]:

```
from scipy.linalg import eigh
nDim = X.shape[0]
corX = (X @ X.T) / nSamples
eigvals, eigvecs = eigh(corX)
V  = np.fliplr(eigvecs)
D = np.zeros(shape=(nDim,nDim))
for i in range(nDim):
    if (eigvals[::-1][i] < 10):
        D[i,i] = 0
    else:
        D[i,i] = np.sqrt(1/eigvals[::-1][i])
Z = V @ D @ V.T @ X
```

Next let's see the transformed data and the new correlation matrix. Figure 3-3 shows that the differentiated features of the character are accentuated and the correlation matrix is diagonal and not distributed along all pixels, showing that the data is whitened with the identity correlation matrix.

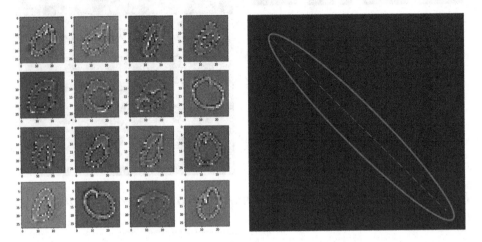

Figure 3-3. *Handwritten MNIST number 0 after whitening. The correlation matrix of all characters is diagonal*

We define the data correlation matrix A as

$$A_k = \frac{1}{k}\sum_{i=1}^{k}\mathbf{x}_i\mathbf{x}_i^T = A_{k-1} + \frac{1}{k}\left(\mathbf{x}_k\mathbf{x}_k^T - A_{k-1}\right).$$

The square root of A, also called the Cholesky decomposition [Cholesky decomposition, Wikipedia], is denoted by $A^{\frac{1}{2}}$. Similarly, $A^{-\frac{1}{2}}$ denotes the inverse square root of A. However, as explained below, there is no unique solution for both of these matrix functions, and various adaptive algorithms can be obtained where each algorithm leads to a different solution. Even though there are numerous solutions for these matrix functions, there are unique solutions under some restrictions, such as a unique symmetric positive definite solution.

In this chapter, I present three adaptive algorithms for each matrix function $A^{\frac{1}{2}}$ and $A^{-\frac{1}{2}}$. Out of the three, one adaptive algorithm leads to the symmetric positive definite solution and two lead to more general solutions.

Various Solutions for $A^{\frac{1}{2}}$ and $A^{-\frac{1}{2}}$

Let $A=\Phi\Lambda\Phi^T$ be the eigen-decomposition of the real, symmetric, positive definite nXn matrix A, where Φ and Λ are respectively the eigenvector and eigenvalue matrices of A. Here $\Lambda=diag(\lambda_1,...,\lambda_n)$ is the diagonal eigenvalue matrix with $\lambda_1\geq...\geq\lambda_n>0$, and $\Phi\in\mathfrak{R}^{nXn}$ is orthonormal. A solution for $A^{\frac{1}{2}}$ is $L=\Phi D$, where $D=diag\left(\pm\lambda_1^{\frac{1}{2}},...,\pm\lambda_n^{\frac{1}{2}}\right)$. However, in general this is not a symmetric solution, and for any orthonormal[1] matrix U, ΦDU is also a solution. We can show that $A^{\frac{1}{2}}$ is symmetric if, and only if, it is of the form $\Phi D\Phi^T$, and there are 2^n symmetric solutions for $A^{\frac{1}{2}}$. When D is positive definite, we obtain the unique symmetric positive definite solution for $A^{\frac{1}{2}}$ as $\Phi\Lambda^{\frac{1}{2}}\Phi^T$, where $\Lambda^{\frac{1}{2}}=diag\left(\lambda_1^{\frac{1}{2}},...,\lambda_n^{\frac{1}{2}}\right)$. Similarly, a general solution for the inverse square root $A^{-\frac{1}{2}}$ of A is $\Phi D^{-1}U$, where D is defined before and U is any orthonormal matrix. The unique symmetric positive definite solution for $A^{-\frac{1}{2}}$ is $\Phi\Lambda^{-\frac{1}{2}}\Phi^T$, where $\Lambda^{-\frac{1}{2}}=diag\left(\lambda_1^{-\frac{1}{2}},...,\lambda_n^{-\frac{1}{2}}\right)$.

Outline of This Chapter

In sections 3.2, 3.3, and 3.4, I discuss three algorithms for the adaptive computation of $A^{\frac{1}{2}}$. Section 3.4 discusses the unique symmetric positive definite solution for $A^{\frac{1}{2}}$. In Sections 3.5, 3.6, and 3.7, I discuss three algorithms for the adaptive computation of $A^{-\frac{1}{2}}$. Section 3.7 describes the unique symmetric positive definite solution for $A^{-\frac{1}{2}}$. Section 3.8 presents experimental results for the six algorithms with 10-dimensional Gaussian data. Section 3.9 concludes the chapter.

---

[1] An orthonormal matrix U has the property $UU^T=U^TU=I$ (identity).

3.2 Adaptive Square Root Algorithm: Method 1

Let $\{\mathbf{x}_k \in \Re^n\}$ be a sequence of data vectors whose online data correlation matrix $A_k \in \Re^{nXn}$ is given by

$$A_k = \frac{1}{k}\sum_{i=1}^{k}\beta^{k-i}\mathbf{x}_i\mathbf{x}_i^T . \tag{3.1}$$

Here $\mathbf{x}_k$ is an observation vector at time k and $0<\beta\le1$ is a forgetting factor used for non-stationary sequences. If the data is stationary, the asymptotic correlation matrix A is

$$A = \lim_{k\to\infty} E[A_k]. \tag{3.2}$$

Objective Function

Following the methodology described in Section 1.4, I present the algorithm by first showing an objective function J, whose minimum with respect to matrix W gives us the square root of the asymptotic data correlation matrix A. The objective function is

$$J(W) = \|A - W^TW\|_F^2 . \tag{3.3}$$

The gradient of $J(W)$ with respect to W is

$$\nabla_W J(W) = -4W(A - W^TW). \tag{3.4}$$

Adaptive Algorithm

From the gradient in (3.4), we obtain the following adaptive gradient descent algorithm:

$$W_{k+1} = W_k - \eta_k(1/4)\nabla_W J(W_k;A_k) = W_k + \eta_k\left(W_kA_k - W_kW_k^TW_k\right), \tag{3.5}$$

where η_k is a small decreasing constant and follows assumption A1.2 in the Proofs of Convergence in the GitHub [Chatterjee GitHub].

The following Python code implements this algorithm with data X[nDim,nSamples]:

```
for epoch in range(nEpochs):
    for iter in range(nSamples):
        cnt = nSamples*epoch + iter
        x = X[:,iter]
        x = x.reshape(nDim,1)
        A = A + (1.0/(1 + cnt))*((np.dot(x, x.T)) - A)
        etat1 = 1.0/(50 + cnt)
        # Algorithm 1
        W1 = W1 + etat1 * (W1 @ A - W1 @ W1.T @ W1)
```

3.3 Adaptive Square Root Algorithm: Method 2

Objective Function

The objective function $J(W)$, whose minimum with respect to W gives us the square root of A, is

$$J(W) = \left\| A - WW^T \right\|_F^2 . \tag{3.6}$$

The gradient of $J(W)$ with respect to W is

$$\nabla_w J(W) = -4(A - WW^T)W. \tag{3.7}$$

Adaptive Algorithm

We obtain the following adaptive gradient descent algorithm for square root of A:

$$W_{k+1} = W_k - \eta_k (1/4) \nabla_W J(W_k; A_k) = W_k + \eta_k \left(A_k W_k - W_k W_k^T W_k \right), \qquad (3.8)$$

where η_k is a small decreasing constant.

The following Python code implements this algorithm with data X[nDim,nSamples]:

```python
for epoch in range(nEpochs):
    for iter in range(nSamples):
        cnt = nSamples*epoch + iter
        x = X[:,iter]
        x = x.reshape(nDim,1)
        A = A + (1.0/(1 + cnt))*((np.dot(x, x.T)) - A)
        etat1 = 1.0/(50 + cnt)
        # Algorithm 2
        W2 = W2 + etat1 * (A @ W2 - W2 @ W2.T @ W2)
```

3.4 Adaptive Square Root Algorithm: Method 3

Adaptive Algorithm

Following the adaptive algorithms (3.5) and (3.8), I now present an algorithm for the computation of a symmetric positive definite square root of A:

$$W_{k+1} = W_k + \eta_k \left(A_k - W_k^2 \right), \qquad (3.9)$$

where η_k is a small decreasing constant and W_k is symmetric.

The following Python code implements this algorithm with data X[nDim,nSamples]:

```
for epoch in range(nEpochs):
    for iter in range(nSamples):
        cnt = nSamples*epoch + iter
        x = X[:,iter]
        x = x.reshape(nDim,1)
        A = A + (1.0/(1 + cnt))*((np.dot(x, x.T)) - A)
        etat2 = 1.0/(50 + cnt)
        # Algorithm 3
        W3 = W3 + etat2 * (A - W3 @ W3)
```

3.5 Adaptive Inverse Square Root Algorithm: Method 1

Objective Function

The objective function $J(W)$, whose minimizer W^* gives us the inverse square root of A, is

$$J(W) = \left\| I - W^T A W \right\|_F^2 . \tag{3.10}$$

The gradient of $J(W)$ with respect to W is

$$\nabla_W J(W) = -4AW(I - W^T AW). \tag{3.11}$$

Adaptive Algorithm

From the gradient in (3.11), we obtain the following adaptive gradient descent algorithm:

$$W_{k+1} = W_k - \eta_k \left(1/4\right) A_k^{-1} \nabla_W J\left(W_k ; A_k\right) = W_k + \eta_k \left(W_k - W_k W_k^T A_k W_k\right) \tag{3.12}$$

The following Python code implements this algorithm with data X[nDim,nSamples]:

```
for epoch in range(nEpochs):
    for iter in range(nSamples):
        cnt = nSamples*epoch + iter
        x = X[:,iter]
        x = x.reshape(nDim,1)
        A = A + (1.0/(1 + cnt))*((np.dot(x, x.T)) - A)
        etat1 = 1.0/(100 + cnt)
        # Algorithm 1
        W1 = W1 + etat1 * (W1 - W1 @ W1.T @ A @ W1)
```

3.6 Adaptive Inverse Square Root Algorithm: Method 2

Objective Function

The objective function $J(W)$, whose minimum with respect to W gives us the inverse square root of A, is

$$J(W) = \|I - WAW^T\|_F^2.$$ (3.13)

The gradient of $J(W)$ with respect to W is

$$\nabla_W J(W) = -4(I - WAW^T)WA.$$ (3.14)

Adaptive Algorithm

We obtain the following adaptive algorithm for the inverse square root of A:

$$W_{k+1} = W_k - \eta_k (1/4)\nabla_W J(W_k; A_k) A_k^{-1} = W_k + \eta_k \left(W_k - W_k A_k W_k^T W_k\right),$$ (3.15)

where η_k is a small decreasing constant.

The following Python code implements this algorithm with data X[nDim,nSamples]:

```python
for epoch in range(nEpochs):
    for iter in range(nSamples):
        cnt = nSamples*epoch + iter
        x = X[:,iter]
        x = x.reshape(nDim,1)
        A = A + (1.0/(1 + cnt))*((np.dot(x, x.T)) - A)
        etat1 = 1.0/(100 + cnt)
        # Algorithm 2
        W2 = W2 + etat1 * (W2 - W2 @ A @ W2.T @ W2)
```

3.7 Adaptive Inverse Square Root Algorithm: Method 3

Adaptive Algorithm

By extending the adaptive algorithms (3.12) and (3.15), I now present an adaptive algorithm for the computation of a symmetric positive definite inverse square root of A:

$$W_{k+1} = W_k + \eta_k(I - W_kA_kW_k), \tag{3.16}$$

where η_k is a small decreasing constant and W_k is symmetric.

The following Python code implements this algorithm with data X[nDim,nSamples]:

```python
for epoch in range(nEpochs):
    for iter in range(nSamples):
        cnt = nSamples*epoch + iter
```

```
x = X[:,iter]
x = x.reshape(nDim,1)
A = A + (1.0/(1 + cnt))*((np.dot(x, x.T)) - A)
etat2 = 1.0/(100 + cnt)
# Algorithm 3
W3 = W3 + etat2 * (I - W3 @ A @ W3)
```

3.8 Experimental Results

I generated 500 samples $\{\mathbf{x}_k\}$ of 10-dimensional (i.e., $n=10$) Gaussian data with mean zero and covariance, given below. The covariance matrix is obtained from the first covariance matrix in [Okada and Tomita 85] multiplied by 3. The covariance matrix is

$$3\begin{bmatrix} 0.091 & 0.038 & -0.053 & -0.005 & 0.010 & -0.136 & 0.155 & 0.030 & 0.002 & 0.032 \\ 0.038 & 0.373 & 0.018 & -0.028 & -0.011 & -0.367 & 0.154 & -0.057 & -0.031 & -0.065 \\ -0.053 & 0.018 & 1.430 & 0.017 & 0.055 & -0.450 & -0.038 & -0.298 & -0.041 & -0.030 \\ -0.005 & -0.028 & 0.017 & 0.084 & -0.005 & 0.016 & 0.042 & -0.022 & 0.001 & 0.005 \\ 0.010 & -0.011 & 0.055 & -0.005 & 0.071 & 0.088 & 0.058 & -0.069 & -0.008 & 0.003 \\ -0.136 & -0.367 & -0.450 & 0.016 & 0.088 & 5.720 & -0.544 & -0.248 & 0.005 & 0.095 \\ 0.155 & 0.154 & -0.038 & 0.042 & 0.058 & -0.544 & 2.750 & -0.343 & -0.011 & -0.120 \\ 0.030 & -0.057 & -0.298 & -0.022 & -0.069 & -0.248 & -0.343 & 1.450 & 0.078 & 0.028 \\ 0.002 & -0.031 & -0.041 & 0.001 & -0.008 & 0.005 & -0.011 & 0.078 & 0.067 & 0.015 \\ 0.032 & -0.065 & -0.030 & 0.005 & 0.003 & 0.095 & -0.120 & 0.028 & 0.015 & 0.341 \end{bmatrix}.$$

The eigenvalues of the covariance matrix are

[17.699, 8.347, 5.126, 3.088, 1.181, 0.882, 0.261, 0.213, 0.182, 0.151].

Experiments for Adaptive Square Root Algorithms

I used adaptive algorithms (3.5), (3.8), and (3.9) for Methods 1, 2, and 3, respectively, to compute the square root A, where A is the correlation matrix computed from all collected samples as

$$A = \frac{1}{500} \sum_{i=1}^{500} \mathbf{x}_i \mathbf{x}_i^T .$$

I started the algorithms with $W_0 = I$ (10X10 identity matrix). At k^{th} update of each algorithm, I computed the Frobenius norm [Frobenius norm, Wikipedia] of the error between the actual correlation matrix A and the square of W_k that is appropriate for each method. I denote this error by e_k as follows:

$$e_k^{\text{Method1}} = \left\| A - W_k^T W_k \right\|_F , \tag{3.17}$$

$$e_k^{\text{Method2}} = \left\| A - W_k W_k^T \right\|_F , \tag{3.18}$$

$$e_k^{\text{Method3-1}} = \left\| A - W_k^2 \right\|_F , \tag{3.19}$$

$$e_k^{\text{Method3-2}} = \left\| \Phi \Lambda^{\frac{1}{2}} \Phi^T - W_k \right\|_F . \tag{3.20}$$

For each algorithm, I generated A_k from $\mathbf{x}_k$ by (2.5) with $\beta=1$. In (3.9) I computed the error between the unique positive definite solution of $A^{\frac{1}{2}}$ and its estimate W_k. Figure 3-4 shows the convergence plots for all three methods by plotting the four errors e_k against k. The final values of e_k after 500 samples were $e_{500} = 0.451$ for Methods 1 and 2, $e_{500} = 0.720$ for Method 3-1 (3.19), and $e_{500} = 0.250$ for Method 3-2 (3.20).

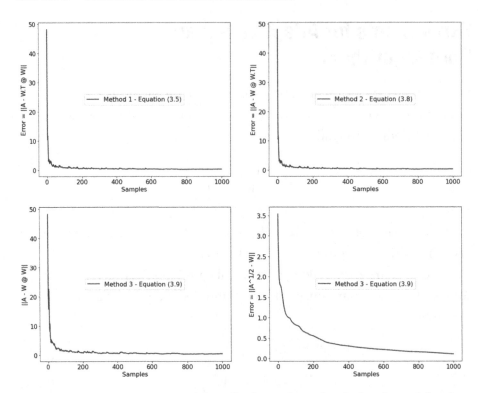

Figure 3-4. *Convergence of the $A^{\frac{1}{2}}$ algorithms (3.5), (3.8), and (3.9)*

It is clear from Figure 3-4 that the errors are all close to zero. The small differences compared to the actual values are due to random fluctuations in the elements of W_k caused by the varying input data.

Experiments for Adaptive Inverse Square Root Algorithms

I used adaptive algorithms (3.12), (3.15), and (3.16) for Methods 1, 2, and 3, respectively, to compute the inverse square root A. I started the algorithms with $W_0 = I$ (10X10 identity matrix). At k^{th} update of each algorithm, I computed the Frobenius norm of the error between the actual correlation matrix A and the inverse square of W_k that is appropriate for each method. I denoted this error by e_k as shown:

$$e_k^{\text{Method1}} = I - W_k^T A W_{kF}, \tag{3.21}$$

$$e_k^{\text{Method2}} = I - W_k A W_{k\ F}^T, \tag{3.22}$$

$$e_k^{\text{Method3 1}} = I - W_k A W_{kF}, \tag{3.23}$$

$$e_k^{\text{Method3 2}} = \Phi \Lambda^{-\frac{1}{2}} \Phi^T - W_{kF}, \tag{3.24}$$

In (3.24) I computed the error between the unique positive definite solution of $A^{-\frac{1}{2}}$ and its estimate W_k. Figure 3-5 shows the convergence plot for all three methods by plotting the four errors e_k against k. The final values of e_k after 500 samples were $e_{500} = 0.419$ for Methods 1 and 2, $e_{500} = 0.630$ for Method 3-1 (3.23), and $e_{500} = 0.823$ for Method 3-2 (3.24).

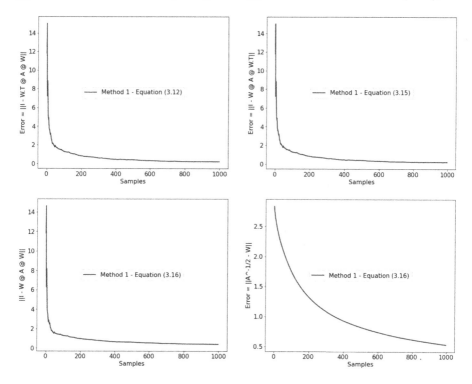

Figure 3-5. *Convergence of the $A^{-\frac{1}{2}}$ algorithms (3.12), (3.15), and (3.16)*

It is clear from Figure 3-5 that the errors are all close to zero. As before, experiments with higher epochs show an improvement in the estimation accuracy.

3.9 Concluding Remarks

I presented six adaptive algorithms for the computation of the square root and inverse square root of the correlation matrix of a random vector sequence. In four cases, I presented an objective function and, in all cases, I discussed the convergence properties of the algorithms. Note that although I applied the gradient descent technique on these objective functions, I could have applied any other technique of nonlinear optimization such as steepest descent, conjugate direction, Newton-Raphson, or recursive least squares. The availability of the objective functions allows us to derive new algorithms by using new optimization techniques on them, and also to perform convergence analyses of the adaptive algorithms.

CHAPTER 4

First Principal Eigenvector

4.1 Introduction and Use Cases

In this chapter, I present a unified framework to derive and discuss ten adaptive algorithms (some well-known) for principal eigenvector computation, which is also known as principal component analysis (PCA) or the Karhunen-Loeve [Karhunen–Loève theorem, Wikipedia] transform. The first *principal eigenvector* of a symmetric positive definite matrix $A \in \Re^{nXn}$ is the eigenvector ϕ_1 corresponding to the largest eigenvalue λ_1 of A. Here $A\phi_i = \lambda_i \phi_i$ for $i=1,...,n$, where $\lambda_1 > \lambda_2 \geq ... \geq \lambda_n > 0$ are the n largest eigenvalues of A corresponding to eigenvectors $\phi_1,...,\phi_n$.

An important problem in machine learning is to extract the most significant feature that represents the variations in the multi-dimensional data. This reduces the multi-dimensional data into one dimension that can be easily modeled. However, in real-world applications, the data statistics change over time (non-stationary). Hence it is challenging to design a solution that adapts to changing data on a low-memory and low-computation edge device.

C. Chatterjee, *Adaptive Machine Learning Algorithms with Python*,
https://doi.org/10.1007/978-1-4842-8017-1_4

Figure 4-1 shows an example of streaming 10-dimensional non-stationary data that abruptly changes statistical properties after 500 samples. The overlaid red curve shows the principal eigenvector estimated by the adaptive algorithm. The adaptive estimate of the principal eigenvector converges to its true value within 50 samples. As the data changes abruptly after 500 samples, it readapts to the changed data and converges back to its true value within 100 samples. All of this is achieved with low computation, low memory, and low latency.

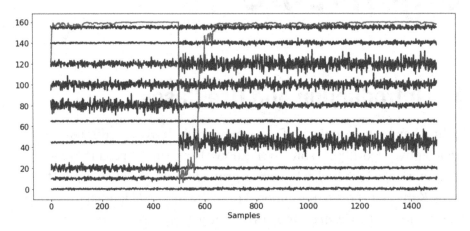

Figure 4-1. *Rapid convergence of the first principal eigenvector is computed by an adaptive algorithm in spite of abrupt changes in data*

Besides this example, there are several applications in machine learning, pattern analysis, signal processing, cellular communications, and automatic control [Haykin 94, Owsley 78, Pisarenko 73, Chatterjee et al. 97-99, Chen et al. 99, Diamantaras and Strintzis 97], where an online (i.e., real-time) solution of principal eigen-decomposition is desired. As discussed in Chapter 2, in these real-time situations, the underlying correlation matrix A is unknown. Instead, we have a sequence of random vectors $\{\mathbf{x}_k \in \Re^n\}$ from which we obtain an instantaneous matrix sequence $\{A_k \in \Re^{n \times n}\}$, such that $A = \lim_{k \to \infty} E[A_k]$. For every incoming sample $\mathbf{x}_k$, we need to obtain the current estimate $\mathbf{w}_k$ of the principal eigenvector $\boldsymbol{\phi}_1$, such that $\mathbf{w}_k$ converges strongly to its true value $\boldsymbol{\phi}_1$.

A common method of computing the online estimate $\mathbf{w}_k$ of $\boldsymbol{\phi}_1$ is to maximize the *Rayleigh quotient* (RQ) [Golub and VanLoan 83] criterion $J(\mathbf{w}_k;A_k)$, where

$$J\left(\mathbf{w}_k;A_k\right) = \frac{\mathbf{w}_k^T A_k \mathbf{w}_k}{\mathbf{w}_k^T \mathbf{w}_k}. \tag{4.1}$$

The signal $\mathbf{x}_k$ can be compressed to a single value by projecting it onto $\mathbf{w}_k$ as $\mathbf{w}_k^T \mathbf{x}_k$.

The literature for PCA algorithms is very diverse and practitioners have approached the problem from a variety of backgrounds including signal processing, neural learning, and statistical pattern recognition. Within each discipline, adaptive PCA algorithms are derived from their own perspectives, which may include ad hoc methods. Since the approaches and solutions to PCA algorithms are distributed along disciplinary lines, a unified framework for deriving and analyzing these algorithms is necessary.

In this chapter, I offer a common framework for derivation, convergence, and rate analyses for the ten adaptive algorithms in four steps outlined in Section 1.4. For each algorithm, I present the results for each of these steps. The unified framework helps in conducting a comparative study of the ten algorithms. In the process, I offer fresh perspectives on known algorithms and present two new adaptive algorithms for PCA. For known algorithms, if results exist from prior implementations, I state them; otherwise, I provide the new results. For the new algorithms, I prove my results.

Outline of This Chapter

In Section 4.2, I list the adaptive PCA algorithms that I derive and discuss in this chapter. I also list the objective functions from which I derive these algorithms and the necessary assumptions. Section 4.3 presents the Oja

PCA algorithm and describes its convergence properties. In Section 4.4, I analyze three algorithms based on the Rayleigh quotient criterion (4.1). In Section 4.5, I discuss PCA algorithms based on the information theoretic criterion. Section 4.6 describes the mean squared error objective function and algorithms. In Section 4.7, I discuss penalty function-based algorithms. Sections 4.8 and 4.9 present new PCA algorithms based on the augmented Lagrangian criteria. Section 4.10 presents the summary of convergence results. Section 4.11 discusses the experimental results. Finally, section 4.12 concludes the chapter.

4.2 Algorithms and Objective Functions
Adaptive Algorithms

[Chatterjee *Neural Networks*, Vol. 18, No. 2, pp. 145-149, March 2005].

I have itemized the algorithms based on their inventors or on the objective functions from which they are derived. All algorithms are of the form

$$\mathbf{w}_{k+1} = \mathbf{w}_k + \eta_k h(\mathbf{w}_k, A_k), \qquad (4.2)$$

where the function $h(\mathbf{w}_k, A_k)$ follows certain continuity and regularity properties [Ljung 77,92], and η_k is a decreasing gain sequence. The term $h(\mathbf{w}_k; A_k)$ for various adaptive algorithms are

- OJA: $A_k \mathbf{w}_k - \mathbf{w}_k \mathbf{w}_k^T A_k \mathbf{w}_k$.

- RQ: $\dfrac{1}{\mathbf{w}_k^T \mathbf{w}_k} \left(A_k \mathbf{w}_k - \mathbf{w}_k \left(\dfrac{\mathbf{w}_k^T A_k \mathbf{w}_k}{\mathbf{w}_k^T \mathbf{w}_k} \right) \right)$.

- OJAN: $A_k \mathbf{w}_k - \mathbf{w}_k \left(\dfrac{\mathbf{w}_k^T A_k \mathbf{w}_k}{\mathbf{w}_k^T \mathbf{w}_k} \right) = \mathbf{w}_k^T \mathbf{w}_k \cdot \text{RQ}$.

- LUO: $\mathbf{w}_k^T \mathbf{w}_k \left(A_k \mathbf{w}_k - \mathbf{w}_k \left(\dfrac{\mathbf{w}_k^T A_k \mathbf{w}_k}{\mathbf{w}_k^T \mathbf{w}_k} \right) \right) = \left(\mathbf{w}_k^T \mathbf{w}_k \right)^2 \cdot \text{RQ}$.

- IT: $\dfrac{A_k \mathbf{w}_k}{\mathbf{w}_k^T A_k \mathbf{w}_k} - \mathbf{w}_k = \dfrac{1}{\mathbf{w}_k^T A_k \mathbf{w}_k} \cdot \text{OJA}.$

- XU: $2 A_k \mathbf{w}_k - \mathbf{w}_k \mathbf{w}_k^T A_k \mathbf{w}_k - A_k \mathbf{w}_k \mathbf{w}_k^T \mathbf{w}_k = \text{OJA} - A_k \mathbf{w}_k \left(\mathbf{w}_k^T \mathbf{w}_k - 1 \right).$

- PF: $A_k \mathbf{w}_k - \mu \mathbf{w}_k \left(\mathbf{w}_k^T \mathbf{w}_k - 1 \right).$

- OJA+: $A_k \mathbf{w}_k - \mathbf{w}_k \mathbf{w}_k^T A_k \mathbf{w}_k - \mathbf{w}_k \left(\mathbf{w}_k^T \mathbf{w}_k - 1 \right) = \text{OJA} - \mathbf{w}_k \left(\mathbf{w}_k^T \mathbf{w}_k - 1 \right)$

- AL1: $A_k \mathbf{w}_k - \mathbf{w}_k \mathbf{w}_k^T A_k \mathbf{w}_k - \mu \mathbf{w}_k \left(\mathbf{w}_k^T \mathbf{w}_k - 1 \right).$

- AL2: $2 A_k \mathbf{w}_k - \mathbf{w}_k \mathbf{w}_k^T A_k \mathbf{w}_k - A_k \mathbf{w}_k \mathbf{w}_k^T \mathbf{w}_k - \mu \mathbf{w}_k \left(\mathbf{w}_k^T \mathbf{w}_k - 1 \right).$

Here IT denotes information theory, and AL denotes augmented Lagrangian. Although most of these algorithms are known, the new AL1 and AL2 algorithms are derived from an augmented Lagrangian objective function discussed later in this chapter.

Objective Functions

Conforming to my proposed methodology in Chapter 2.2, all algorithms mentioned before are derived from objective functions. Some of these objective functions are

- Objective function for the OJA algorithm,

- Least mean squared error criterion,

- Rayleigh quotient criterion,

- Penalty function method,

- Information theory criterion, and

- Augmented Lagrangian method.

4.3 OJA Algorithm

This algorithm was given by Oja et al. [Oja 85, 89, 92]. Intuitively, the OJA algorithm is derived from the Rayleigh quotient criterion by representing it as a Lagrange function, which minimizes $-\mathbf{w}_k^T A_k \mathbf{w}_k$ under the constraint $\mathbf{w}_k^T \mathbf{w}_k = 1$.

Objective Function

In terms of the data samples $\mathbf{x}_k$, the objective function for the OJA algorithm can be written as

$$J\left(\mathbf{w}_k ; \mathbf{x}_k\right) = -\left\| \mathbf{x}_k^T \left(\mathbf{x}_k - \mathbf{w}_k \mathbf{w}_k^T \mathbf{x}_k \right) \right\|^2. \tag{4.3}$$

If we represent the data correlation matrix A_k by its instantaneous value $\mathbf{x}_k \mathbf{x}_k^T$, then (4.3) is equivalent to the following objective function:

$$J\left(\mathbf{w}_k ; A_k\right) = -\left\| A_k^{\frac{1}{2}} \left(I - \mathbf{w}_k \mathbf{w}_k^T \right) A_k^{\frac{1}{2}} \right\|_F^2. \tag{4.4}$$

We see from (4.4) that the objective function $J(\mathbf{w}_k;\mathbf{x}_k)$ represents the difference between the sample $\mathbf{x}_k$ and its transformation due to a matrix $\mathbf{w}_k \mathbf{w}_k^T$. In neural networks, this transform is called *auto-association*[1] [Haykin 94]. Figure 4-2 shows a two-layer auto-associative network.

---

[1] In the auto-associative mode, the output of the network is desired to be same as the input.

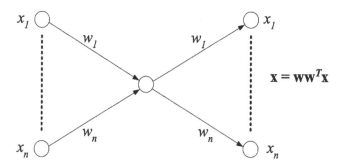

Figure 4-2. *Two-layer linear auto-associative neural network for the first principal eigenvector*

Adaptive Algorithm

The gradient of (4.4) with respect to $\mathbf{w}_k$ is

$$\nabla_{\mathbf{w}_k} J\left(\mathbf{w}_k; A_k\right) = -4A_k\left(A_k\mathbf{w}_k - \mathbf{w}_k\mathbf{w}_k^T A_k\mathbf{w}_k\right).$$

The adaptive gradient descent OJA algorithm for PCA is

$$\mathbf{w}_{k+1} = \mathbf{w}_k - \eta_k A_k^{-1}\nabla_{\mathbf{w}_k} J\left(\mathbf{w}_k; A_k\right) = \mathbf{w}_k + \eta_k\left(A_k\mathbf{w}_k - \mathbf{w}_k\mathbf{w}_k^T A_k\mathbf{w}_k\right), \quad (4.5)$$

where η_k is a small decreasing constant.

The Python code for this algorithm with multidimensional data `X[nDim,nSamples]` is

```python
A = np.zeros(shape=(nDim,nDim)) # stores adaptive
correlation matrix
w = 0.1 * np.ones(shape=(nDim,11)) # weight vectors of all
algorithms
for epoch in range(nEpochs):
    for iter in range(nSamples):
        cnt = nSamples*epoch + iter
        x = X[:,iter]
        x = x.reshape(nDim,1)
```

```
A = A + (1.0/(1 + cnt))*((np.dot(x, x.T)) - A)
# OJA Algorithm
v = w[:,0].reshape(nDim, 1)
v = v + (1/(100+cnt))*(A @ v - v @ (v.T @ A @ v))
w[:,0] = v.reshape(nDim)
```

Rate of Convergence

The convergence time constant for the principal eigenvector ϕ_1 is $1/\lambda_1$ and for the minor eigenvectors ϕ_i is $1/(\lambda_1-\lambda_i)$ for $i=2,...,n$. The time constants are dependent on the eigen-structure of the data correlation matrix A.

4.4 RQ, OJAN, and LUO Algorithms

Objective Function

These three algorithms are different derivations of the following Rayleigh quotient objective function:

$$J\left(\mathbf{w}_k;A_k\right) = -\left(\frac{\mathbf{w}_k^T A_k \mathbf{w}_k}{\mathbf{w}_k^T \mathbf{w}_k}\right). \tag{4.6}$$

These algorithms were initially presented by Luo et al. [Luo et al. 97; Taleb et al. 99; Cirrincione et al. 00] and Oja et al. [Oja et al. 92]. Variations of the RQ algorithm have been presented by many practitioners [Chauvin 89; Sarkar et al. 89; Yang et al. 89; Fu and Dowling 95; Taleb et al. 99; Cirrincione et al. 00].

Adaptive Algorithms

The gradient of (4.7) with respect to $\mathbf{w}_k$ is

$$\nabla_{\mathbf{w}_k} J\left(\mathbf{w}_k; A_k\right) = \frac{-1}{\mathbf{w}_k^T \mathbf{w}_k}\left(A_k \mathbf{w}_k - \mathbf{w}_k \frac{\mathbf{w}_k^T A_k \mathbf{w}_k}{\mathbf{w}_k^T \mathbf{w}_k}\right).$$

The adaptive gradient descent RQ algorithm for PCA is

$$\mathbf{w}_{k+1} = \mathbf{w}_k - \eta_k \nabla_{\mathbf{w}_k} J\left(\mathbf{w}_k; A_k\right) = \mathbf{w}_k + \eta_k \frac{1}{\mathbf{w}_k^T \mathbf{w}_k}\left(A_k \mathbf{w}_k - \mathbf{w}_k \frac{\mathbf{w}_k^T A_k \mathbf{w}_k}{\mathbf{w}_k^T \mathbf{w}_k}\right). \quad (4.7)$$

The adaptive gradient descent OJAN algorithm for PCA is

$$\mathbf{w}_{k+1} = \mathbf{w}_k - \eta_k \left(\mathbf{w}_k^T \mathbf{w}_k\right) \nabla_{\mathbf{w}_k} J\left(\mathbf{w}_k; A_k\right) = \mathbf{w}_k + \eta_k \left(A_k \mathbf{w}_k - \mathbf{w}_k \frac{\mathbf{w}_k^T A_k \mathbf{w}_k}{\mathbf{w}_k^T \mathbf{w}_k}\right). \quad (4.8)$$

The adaptive gradient descent LUO algorithm for PCA is

$$\begin{aligned}
\mathbf{w}_{k+1} &= \mathbf{w}_k - \eta_k \left(\mathbf{w}_k^T \mathbf{w}_k\right)^2 \nabla_{\mathbf{w}_k} J\left(\mathbf{w}_k; A_k\right) \\
&= \mathbf{w}_k + \eta_k \left(\mathbf{w}_k^T \mathbf{w}_k\right)\left(A_k \mathbf{w}_k - \mathbf{w}_k \frac{\mathbf{w}_k^T A_k \mathbf{w}_k}{\mathbf{w}_k^T \mathbf{w}_k}\right).
\end{aligned} \quad (4.9)$$

The Python code for these algorithms with multidimensional data X[nDim,nSamples] is

```python
A = np.zeros(shape=(nDim,nDim)) # stores adaptive
correlation matrix
w = 0.1 * np.ones(shape=(nDim,11)) # weight vectors of all
algorithms
for epoch in range(nEpochs):
    for iter in range(nSamples):
        cnt = nSamples*epoch + iter
        x = X[:,iter]
        x = x.reshape(nDim,1)
        A = A + (1.0/(1 + cnt))*((np.dot(x, x.T)) - A)
        # OJAN Algorithm
```

```
v = w[:,1].reshape(nDim, 1)
v = v + (1/(10+cnt))*(A @ v - v @ ((v.T @ A @ v) /
(v.T @ v)) )
w[:,1] = v.reshape(nDim)
# LUO Algorithm
v = w[:,2].reshape(nDim, 1)
v = v + (1/(20+cnt))*(A @ v * (v.T @ v) - v @ (v.T
@ A @ v))
w[:,2] = v.reshape(nDim)
# RQ Algorithm
v = w[:,3].reshape(nDim, 1)
v = v + (1/(100+cnt))*(A @ v - v @ ((v.T @ A @ v) /
(v.T @ v)) )
w[:,3] = v.reshape(nDim)
```

Rate of Convergence

The convergence time constants for *principal eigenvector* ϕ_1 are

RQ: $\|\mathbf{w}_0\|^2/\lambda_1$.
OJAN: $1/\lambda_1$.
LUO: $\|\mathbf{w}_0\|^{-2}/\lambda_1$.

The convergence time constants for the minor eigenvectors ϕ_i $(i=2,\ldots,n)$ are

RQ: $\|\mathbf{w}_0\|^2/(\lambda_1-\lambda_i)$ for $i=2,\ldots,n$.
OJAN: $1/(\lambda_1-\lambda_i)$ for $i=2,\ldots,n$.
LUO: $\|\mathbf{w}_0\|^{-2}/(\lambda_1-\lambda_i)$ for $i=2,\ldots,n$.

The time constants are dependent on the eigen-structure of A.

4.5 IT Algorithm

Objective Function

The objective function for the information theory (IT) criterion is

$$J\left(\mathbf{w}_k; A_k\right) = \mathbf{w}_k^T \mathbf{w}_k - \ln\left(\mathbf{w}_k^T A_k \mathbf{w}_k\right). \tag{4.10}$$

Plumbley [Pumbley 95] and Miao and Hua [Miao and Hua 98] have studied this objective function.

Adaptive Algorithm

The gradient of (4.12) with respect to $\mathbf{w}_k$ is

$$\nabla_{\mathbf{w}_k} J\left(\mathbf{w}_k; A_k\right) = \mathbf{w}_k - \frac{A_k \mathbf{w}_k}{\mathbf{w}_k^T A_k \mathbf{w}_k}.$$

The adaptive gradient descent IT algorithm for PCA is

$$\mathbf{w}_{k+1} = \mathbf{w}_k - \eta_k \nabla_{\mathbf{w}_k} J\left(\mathbf{w}_k; A_k\right) = \mathbf{w}_k + \eta_k \left(\frac{A_k \mathbf{w}_k}{\mathbf{w}_k^T A_k \mathbf{w}_k} - \mathbf{w}_k\right). \tag{4.11}$$

The Python code for this algorithm with multidimensional data X[nDim,nSamples] is

```
A = np.zeros(shape=(nDim,nDim)) # stores adaptive
correlation matrix
w = 0.1 * np.ones(shape=(nDim,11)) # weight vectors of all
algorithms
for epoch in range(nEpochs):
    for iter in range(nSamples):
        cnt = nSamples*epoch + iter
        x = X[:,iter]
        x = x.reshape(nDim,1)
        A = A + (1.0/(1 + cnt))*((np.dot(x, x.T)) - A)
```

```
# IT Algorithm
v = w[:,5].reshape(nDim, 1)
v = v + (4/(1+cnt))*((A @ v / (v.T @ A @ v)) - v)
w[:,5] = v.reshape(nDim)
```

Rate of Convergence

A unique feature of this algorithm is that the time constant for $\|\mathbf{w}(t)\|$ is 1 and it is *independent* of the eigen-structure of A.

Upper Bound of η_k

I have proven that there exists a uniform upper bound for η_k such that $\mathbf{w}_k$ is uniformly bounded. Furthermore, if $\|\mathbf{w}_k\|^2 \le \alpha+1$, then $\|\mathbf{w}_{k+1}\|^2 \le \|\mathbf{w}_k\|^2$ if

$$\eta_k < \frac{2(\alpha+1)}{\alpha}.$$

4.6 XU Algorithm

Objective Function

Originally presented by Xu [Xu 91, 93], the objective function for XU algorithm is

$$J\left(\mathbf{w}_k; A_k\right) = -\mathbf{w}_k^T A_k \mathbf{w}_k + \mathbf{w}_k^T A_k \mathbf{w}_k \left(\mathbf{w}_k^T \mathbf{w} - 1\right)$$
$$= -2\mathbf{w}_k^T A_k \mathbf{w}_k + \mathbf{w}_k^T A_k \mathbf{w}_k \mathbf{w}_k^T \mathbf{w}_k. \tag{4.12}$$

The objective function $J(\mathbf{w}_k; A_k)$ represents the mean squared error between the sample $\mathbf{x}_k$ and its transformation due to a matrix $\mathbf{w}_k \mathbf{w}_k^T$. This transform, also known as *auto-association*, is shown in Figure 4-1. We define $A_k = (1/k)\sum_{t=1}^{k} \mathbf{x}_t \mathbf{x}_t^T$. Then, the mean squared error objective function is

$$J\left(\mathbf{w}_k;A_k\right)=\frac{1}{k}\sum_{i=1}^{k}\left\|\mathbf{x}_k-\mathbf{w}_k\mathbf{w}_k^T\mathbf{x}_k\right\|^2=trA_k-2\mathbf{w}_k^TA_k\mathbf{w}_k+\mathbf{w}_k^TA_k\mathbf{w}_k\mathbf{w}_k^T\mathbf{w}_k,$$

which is the same as (4.12).

Adaptive Algorithm

The gradient of (4.12) with respect to $\mathbf{w}_k$ is

$$\nabla_{\mathbf{w}_k}J\left(\mathbf{w}_k;A_k\right)=-\left(2A_k\mathbf{w}_k-\mathbf{w}_k\mathbf{w}_k^TA_k\mathbf{w}_k-A_k\mathbf{w}_k\mathbf{w}_k^T\mathbf{w}_k\right).$$

The adaptive gradient descent XU algorithm for PCA is

$$\begin{aligned}\mathbf{w}_{k+1}&=\mathbf{w}_k-\eta_k\nabla_{\mathbf{w}_k}J\left(\mathbf{w}_k;A_k\right)\\&=\mathbf{w}_k+\eta_k\left(2A_k\mathbf{w}_k-\mathbf{w}_k\mathbf{w}_k^TA_k\mathbf{w}_k-A_k\mathbf{w}_k\mathbf{w}_k^T\mathbf{w}_k\right).\end{aligned}\tag{4.13}$$

The Python code for this algorithm with multidimensional data X[nDim,nSamples] is

```
A = np.zeros(shape=(nDim,nDim)) # stores adaptive
correlation matrix
w = 0.1 * np.ones(shape=(nDim,11)) # weight vectors of all
algorithms
for epoch in range(nEpochs):
    for iter in range(nSamples):
        cnt = nSamples*epoch + iter
        x = X[:,iter]
        x = x.reshape(nDim,1)
        A = A + (1.0/(1 + cnt))*((np.dot(x, x.T)) - A)
        # XU Algorithm
        v = w[:,6].reshape(nDim, 1)
        v = v + (1/(50+cnt))*(2*A@ v - v@(v.T @ A @ v) - A@ v@
        (v.T @ v))
        w[:,6] = v.reshape(nDim)
```

Rate of Convergence

The convergence time constant for the principal eigenvector $\boldsymbol{\phi}_1$ is $1/\lambda_1$ and for the minor eigenvectors $\boldsymbol{\phi}_i$ is $1/(\lambda_1-\lambda_i)$ for $i=2,\ldots,n$. The time constants are dependent on the eigen-structure of the data correlation matrix A.

Upper Bound of η_k

There exists a uniform upper bound for η_k such that $\mathbf{w}_k$ is uniformly bounded w.p.1. If $\|\mathbf{w}_k\|^2 \leq \alpha+1$ and θ is the largest eigenvalue of A_k, then $\|\mathbf{w}_{k+1}\|^2 \leq \|\mathbf{w}_k\|^2$ if

$$\eta_k < \frac{1}{\theta\alpha}.$$

4.7 Penalty Function Algorithm

Objective Function

Originally given by Chauvin [Chauvin 89], the objective function for the penalty function (PF) algorithm is

$$J\left(\mathbf{w}_k;A_k\right) = -\mathbf{w}_k^T A_k \mathbf{w}_k + \frac{\mu}{2}\left(\mathbf{w}_k^T\mathbf{w}_k - 1\right)^2, \mu > 0. \tag{4.14}$$

The objective function $J(\mathbf{w}_k;A_k)$ is an implementation of the Rayleigh quotient criterion (4.1), where the constraint $\mathbf{w}_k^T\mathbf{w}_k = 1$ is enforced by the penalty function method of nonlinear optimization, and μ is a positive penalty constant.

Adaptive Algorithm

The gradient of (4.14) with respect to $\mathbf{w}_k$ is

$$(1/2)\nabla_{\mathbf{w}_k} J\left(\mathbf{w}_k; A_k\right) = -\left(A_k \mathbf{w}_k - \mu \mathbf{w}_k \left(\mathbf{w}_k^T \mathbf{w}_k - 1\right)\right).$$

The adaptive gradient descent PF algorithm for PCA is

$$\mathbf{w}_{k+1} = \mathbf{w}_k - \eta_k \nabla_{\mathbf{w}_k} J\left(\mathbf{w}_k; A_k\right) = \mathbf{w}_k + \eta_k \left(A_k \mathbf{w}_k - \mu \mathbf{w}_k \left(\mathbf{w}_k^T \mathbf{w}_k - 1\right)\right), \quad (4.15)$$

where $\mu > 0$.

The Python code for this algorithm with multidimensional data
X[nDim,nSamples] is

```
mu = 10
A = np.zeros(shape=(nDim,nDim)) # stores adaptive
correlation matrix
w = 0.1 * np.ones(shape=(nDim,11)) # weight vectors of all
algorithms
for epoch in range(nEpochs):
    for iter in range(nSamples):
        cnt = nSamples*epoch + iter
        x = X[:,iter]
        x = x.reshape(nDim,1)
        A = A + (1.0/(1 + cnt))*((np.dot(x, x.T)) - A)
        # PF Algorithm
        v = w[:,7].reshape(nDim, 1)
        v = v + (1/(50+cnt)) * (A @ v - mu * v @ (v.T @ v - 1))
        w[:,7] = v.reshape(nDim)
```

Rate of Convergence

The convergence time constant for the principal eigenvector $\boldsymbol{\phi}_1$ is $1/(\lambda_1 + \mu)$ and for the minor eigenvectors $\boldsymbol{\phi}_i$ is $1/(\lambda_1 - \lambda_i)$ for $i=2,...,n$. The time constants are dependent on the eigen-structure of the data correlation matrix A.

Upper Bound of η_k

Then there exists a uniform upper bound for η_k such that $\mathbf{w}_k$ is uniformly bounded. If $\|\mathbf{w}_k\|^2 \leq \alpha+1$ and θ is the largest eigenvalue of A_k, then $\|\mathbf{w}_{k+1}\|^2 \leq \|\mathbf{w}_k\|^2$ if

$$\eta_k < \frac{1}{\mu\alpha - \theta}, \text{ assuming } \mu\alpha > \theta.$$

4.8 Augmented Lagrangian 1 Algorithm

Objective Function and Adaptive Algorithm

The objective function for the augmented Lagrangian 1 (AL1) algorithm is obtained by applying the augmented Lagrangian method of nonlinear optimization to minimize $-\mathbf{w}_k^T A_k \mathbf{w}_k$ under the constraint $\mathbf{w}_k^T \mathbf{w}_k = 1$:

$$J\left(\mathbf{w}_k; A_k\right) = -\mathbf{w}_k^T A_k \mathbf{w}_k + \alpha_k \left(\mathbf{w}_k^T \mathbf{w}_k - 1\right) + \frac{\mu}{2}\left(\mathbf{w}_k^T \mathbf{w}_k - 1\right)^2, \tag{4.16}$$

where α_k is a Lagrange multiplier and μ is a positive penalty constant. The gradient of $J(\mathbf{w}_k; A_k)$ with respect to $\mathbf{w}_k$ is

$$\nabla_{\mathbf{w}_k} J\left(\mathbf{w}_k; A_k\right) = -2\left(A_k \mathbf{w}_k - \alpha_k \mathbf{w}_k - \mu\mathbf{w}_k\left(\mathbf{w}_k^T \mathbf{w}_k - 1\right)\right).$$

By equating the gradient to $\mathbf{0}$ and using the constraint $\mathbf{w}_k^T \mathbf{w}_k = 1$, we obtain $\alpha_k = \mathbf{w}_k^T A_k \mathbf{w}_k$. Replacing this α_k in the gradient, we obtain the AL1 algorithm

$$\mathbf{w}_{k+1} = \mathbf{w}_k + \eta_k \left(A_k \mathbf{w}_k - \mathbf{w}_k \mathbf{w}_k^T A_k \mathbf{w}_k - \mu \mathbf{w}_k \left(\mathbf{w}_k^T \mathbf{w}_k - 1 \right) \right), \tag{4.17}$$

where $\mu > 0$. Note that (4.17) is the same as OJA+ algorithm for $\mu = 1$.

The Python code for this algorithm with multidimensional data X[nDim,nSamples] is

```
mu = 10
A = np.zeros(shape=(nDim,nDim)) # stores adaptive
correlation matrix
w = 0.1 * np.ones(shape=(nDim,11)) # weight vectors of all
algorithms
for epoch in range(nEpochs):
    for iter in range(nSamples):
        cnt = nSamples*epoch + iter
        x = X[:,iter]
        x = x.reshape(nDim,1)
        A = A + (1.0/(1 + cnt))*((np.dot(x, x.T)) - A)
        # AL1 Algorithm
        v = w[:,8].reshape(nDim, 1)
        v = v + (1/(50+cnt))*(A@v - v@(v.T @A @ v) - mu* v@
        (v.T@v - 1))
        w[:,8] = v.reshape(nDim)
```

Rate of Convergence

The convergence time constant for the principal eigenvector ϕ_1 is $1/(\lambda_1 + \mu)$ and for the minor eigenvectors ϕ_i is $1/(\lambda_1 - \lambda_i)$ for $i=2,...,n$. The time constants are dependent on the eigen-structure of the data correlation matrix A.

Upper Bound of η_k

There exists a uniform upper bound for η_k such that $\mathbf{w}_k$ is uniformly bounded. If $\|\mathbf{w}_k\|^2 \leq \alpha+1$ and θ is the largest eigenvalue of A_k, then $\|\mathbf{w}_{k+1}\|^2 \leq \|\mathbf{w}_k\|^2$ if

$$\eta_k < \frac{1}{(\mu+\theta)\alpha}.$$

4.9 Augmented Lagrangian 2 Algorithm

Objective Function

The objective function for the augmented Lagrangian 2 (AL2) algorithm is

$$J\left(\mathbf{w}_k;A_k\right) = -\mathbf{w}_k^T A_k \mathbf{w}_k + \mathbf{w}_k^T A_k \mathbf{w}_k \left(\mathbf{w}_k^T \mathbf{w}_k - 1\right)$$

$$+ \frac{\mu}{2}\left(\mathbf{w}_k^T \mathbf{w}_k - 1\right)^2, \mu > 0. \tag{4.18}$$

The objective function $J(\mathbf{w}_k;A_k)$ is an application of the augmented Lagrangian method on the Rayleigh quotient criterion (4.1). It uses the XU objective function and also uses the penalty function method (4.14), where μ is a positive penalty constant.

Adaptive Algorithm

The gradient of (4.18) with respect to $\mathbf{w}_k$ is

$$(1/2)\nabla_{\mathbf{w}_k} J\left(\mathbf{w}_k;A_k\right) = -\left(2A_k\mathbf{w}_k - \mathbf{w}_k\mathbf{w}_k^T A_k\mathbf{w}_k - A_k\mathbf{w}_k\mathbf{w}_k^T\mathbf{w}_k - \mu\mathbf{w}_k\left(\mathbf{w}_k^T\mathbf{w}_k - 1\right)\right).$$

The adaptive gradient descent AL2 algorithm for PCA is

$$\mathbf{w}_{k+1} = \mathbf{w}_k + \eta_k \left(2A_k\mathbf{w}_k - \mathbf{w}_k\mathbf{w}_k^T A_k\mathbf{w}_k - A_k\mathbf{w}_k\mathbf{w}_k^T\mathbf{w}_k - \mu\mathbf{w}_k \left(\mathbf{w}_k^T\mathbf{w}_k - 1 \right) \right), \quad (4.19)$$

where $\mu > 0$.

The Python code for this algorithm with multidimensional data X[nDim,nSamples] is

```python
mu = 10
A = np.zeros(shape=(nDim,nDim)) # stores adaptive
correlation matrix
w = 0.1 * np.ones(shape=(nDim,11)) # weight vectors of all
algorithms
for epoch in range(nEpochs):
    for iter in range(nSamples):
        cnt = nSamples*epoch + iter
        x = X[:,iter]
        x = x.reshape(nDim,1)
        A = A + (1.0/(1 + cnt))*((np.dot(x, x.T)) - A)
        # AL2 Algorithm
        v = w[:,9].reshape(nDim, 1)
        v = v + (1/(50+cnt))*(2* A @ v - v @ (v.T @ A @ v) -
                              A@ v @ (v.T @ v) - mu* v @ (v.T
                              @ v -1))
                                        w[:,9] = v.reshape(nDim)
```

Rate of Convergence

The convergence time constant for the principal eigenvector $\boldsymbol{\phi}_1$ is $1/(\lambda_1 + (\mu/2))$ and for the minor eigenvectors $\boldsymbol{\phi}_i$ is $1/(\lambda_1-\lambda_i)$ for $i=2,...,n$. The time constants are dependent on the eigen-structure of the data correlation matrix A.

Upper Bound of η_k

There exists a uniform upper bound for η_k such that $\mathbf{w}_k$ is uniformly bounded. Furthermore, if $\|\mathbf{w}_k\|^2 \leq \alpha+1$ and θ is the largest eigenvalue of A_k, then $\|\mathbf{w}_{k+1}\|^2 \leq \|\mathbf{w}_k\|^2$ if

$$\eta_k < \frac{2}{(2\theta + \mu)\alpha}.$$

4.10 Summary of Algorithms

Table 4-1 summarizes the convergence results of the algorithms. It also shows the upper bounds of η_k, when available. Here τ denotes the time constant, $\mathbf{w}_0$ denotes the initial value of $\mathbf{w}_k$, $\alpha+1$ the upper bound of $\|\mathbf{w}_k\|^2$ (i.e., $\|\mathbf{w}_k\|^2 \leq \alpha+1$), and θ denotes the first principal eigenvalue of A_k.

Table 4-1. *Summary of Convergence Results*

Algorithm	Convergence Time Constants	Upper Bounds of η_k
OJA	$1/\lambda_1$	$2/\alpha\theta$
OJAN	$1/\lambda_1$	Not Available
LUO	$\|\mathbf{w}_0\|^{-2}/\lambda_1$	Not Available
RQ	$\|\mathbf{w}_0\|^2/\lambda_1$	Not Available
OJA+	$1/(\lambda_1+1)$	$1/(\alpha-\theta)$
IT	1	$2(\alpha+1)/\alpha$
XU	$1/\lambda_1$	$1/\alpha\theta$
PF	$1/(\lambda_1+\mu)$	$1/(\mu\alpha-\theta)$
AL1	$1/(\lambda_1+\mu)$	$1/(\mu+\theta)\alpha$
AL2	$1/(\lambda_1+(\mu/2))$	$2/(\mu+2\theta)\alpha$

Note that a smaller time constant yields faster convergence. The conclusions are

1. For all algorithms, except IT, convergence of $\boldsymbol{\phi}_1$ improves for larger values of λ_1.

2. For LUO, the time constant decreases for larger $\|\mathbf{w}_0\|$, which implies that convergence improves for larger initial weights.

3. For RQ, convergence deteriorates for larger $\|\mathbf{w}_0\|$.

4. For the PF, AL1, and AL2 algorithms, the time constant decreases for larger μ, although very large values of μ will make the algorithm perform poorly due excessive emphasis on the constraints.

4.11 Experimental Results

I did three sets of experiments.

1. In the first experiment, I used the adaptive algorithms described before on a single data set with various starting vectors $\mathbf{w}_0$ [Chatterjee, *Neural Networks*, Vol. 18, No. 2, pp. 145-149, March 2005].

2. In the second experiment, I generated several data samples and used the adaptive algorithms with the same starting vector $\mathbf{w}_0$ [Chatterjee, *Neural Networks*, Vol. 18, No. 2, pp. 145-149, March 2005].

3. In the third experiment, I used a **real-world non-stationary** data set from a public dataset [V. Souza et al. 2020] to demonstrate the fast convergence of the adaptive algorithms to the first principal eigenvector of the ensemble correlation matrix.

Experiments with Various Starting Vectors w_0

[Chatterjee, *Neural Networks*, Vol. 18, No. 2, pp. 145-149, March 2005].

I generated 1,000 samples of 10-dimensional Gaussian data (i.e., $n=10$) with the mean zero and covariance given below. The covariance matrix is obtained from the second covariance matrix in [Okada and Tomita 85] multiplied by 3, which is

$$3\begin{bmatrix} 0.427 & 0.011 & -0.005 & -0.025 & 0.089 & -0.079 & -0.019 & 0.074 & 0.089 & 0.005 \\ 0.011 & 5.690 & -0.069 & -0.282 & -0.731 & 0.090 & -0.124 & 0.100 & 0.432 & -0.103 \\ -0.005 & -0.069 & 0.080 & 0.098 & 0.045 & -0.041 & 0.023 & 0.022 & -0.035 & 0.012 \\ -0.025 & -0.282 & 0.098 & 2.800 & -0.107 & 0.150 & -0.193 & 0.095 & -0.226 & 0.046 \\ 0.089 & -0.731 & 0.045 & -0.107 & 3.440 & 0.253 & 0.251 & 0.316 & 0.039 & -0.010 \\ -0.079 & 0.090 & -0.041 & 0.150 & 0.253 & 2.270 & -0.180 & 0.295 & -0.039 & -0.113 \\ -0.019 & -0.124 & 0.023 & -0.193 & 0.251 & -0.180 & 0.327 & 0.027 & 0.026 & -0.016 \\ 0.074 & 0.100 & 0.022 & 0.095 & 0.316 & 0.295 & 0.027 & 0.727 & -0.096 & -0.017 \\ 0.089 & 0.432 & -0.035 & -0.226 & 0.039 & -0.039 & 0.026 & -0.096 & 0.715 & -0.009 \\ 0.005 & -0.103 & 0.012 & 0.046 & -0.010 & -0.113 & -0.016 & -0.017 & -0.009 & 0.065 \end{bmatrix}. \qquad (4.20)$$

The eigenvalues of the covariance matrix are

17.9013, 10.2212, 8.6078, 6.5361, 2.2396, 1.8369, 1.1361, 0.7693, 0.2245, 0.1503.

I computed the principal eigenvector (i.e., the eigenvector corresponding to the largest eigenvalue = 17.9013) by the adaptive algorithms described before from different starting vectors w_0. I obtained $w_0 = c^* r$, where $r \in \Re^{10}$ is a $N(0,1)$ random vector and $c \in [0.05, 2.0]$. This causes a variation in $\|w_0\|$ from 0.1 to 5.0.

In order to compute the online data sequence $\{A_k\}$, I generated random data vectors $\{\mathbf{x}_k\}$ from the covariance matrix (4.20). I generated $\{A_k\}$ from $\{\mathbf{x}_k\}$ by using the algorithm (2.5) with $\beta=1$. I compute the correlation matrix A after collecting all 500 samples $\mathbf{x}_k$ as

$$A = \frac{1}{500} \sum_{i=1}^{500} \mathbf{x}_i \mathbf{x}_i^T .$$

I refer to the eigenvectors and eigenvalues computed from this A by a standard numerical analysis method [Golub and VanLoan 83] as the *actual values.*

In order to measure the convergence and accuracy of the algorithms, I computed the percentage direction cosine at k^{th} update of each adaptive algorithm as

$$\text{Percentage Direction Cosine } (k) = \frac{100 \left| \mathbf{w}_k^T \boldsymbol{\varphi}_1 \right|}{\left\| \mathbf{w}_k \right\|} , \tag{4.21}$$

where $\mathbf{w}_k$ is the estimated first principal eigenvector of A_k at k^{th} update and $\boldsymbol{\varphi}_1$ is the actual first principal eigenvector computed from all collected samples by a conventional numerical analysis method. For all algorithms, I used $\eta_k=1/(200+k)$. For the PF, AL1, and AL2 algorithms, I used $\mu=10$. The results are summarized in Table 4-2. I reported the percentage direction cosines after sample values $k=N/2$ and N (i.e., $k=250$ and 500) for each algorithm.

Table 4-2. *Convergence of the Principal Eigenvector of A by Adaptive Algorithms at Sample Values* $k=\{250,500\}$ *for Different Initial Values* $\mathbf{w}_0$

$\|\mathbf{w}_0\|$	k	OJA	OJAN	LUO	RQ	OJA+	IT	XU	PF	AL1	AL2
0.1355	250	97.18	97.18	60.78	98.44	97.18	84.53	97.22	97.17	97.18	97.20
	500	99.58	99.58	63.15	99.96	99.58	89.67	99.58	99.58	99.58	99.58
0.4065	250	97.18	97.18	82.44	98.54	97.18	84.96	97.18	97.18	97.18	97.16
	500	99.58	99.58	90.88	99.96	99.58	90.35	99.58	99.58	99.58	99.58
0.6776	250	97.18	97.18	94.63	97.85	97.18	82.55	97.17	97.18	97.18	97.15
	500	99.58	99.58	98.50	99.88	99.58	88.85	99.58	99.58	99.58	99.58
0.9486	250	97.18	97.18	97.05	97.28	97.18	79.60	97.18	97.18	97.18	97.17
	500	99.58	99.58	99.52	99.63	99.58	86.90	99.58	99.58	99.58	99.58
1.2196	250	97.18	97.18	97.60	96.35	97.18	76.67	97.21	97.18	97.18	97.21
	500	99.58	99.58	99.80	99.19	99.58	84.80	99.58	99.58	99.58	99.58
1.4906	250	97.18	97.18	97.97	94.43	97.18	73.99	97.26	97.18	97.17	97.27
	500	99.58	99.58	99.90	98.41	99.58	82.68	99.59	99.58	99.58	99.59
1.7617	250	97.17	97.18	98.31	91.53	97.18	71.63	97.33	97.18	97.16	97.35
	500	99.58	99.58	99.95	97.08	99.58	80.61	99.59	99.58	99.58	99.59

2.0327	250	97.17	97.18	98.57	88.04	97.17	69.61	97.44	97.17	97.15	97.51
	500	99.58	99.58	99.96	95.08	99.58	78.63	99.60	99.58	99.58	99.60
2.3037	250	97.17	97.18	98.75	84.43	97.17	67.90	97.62	97.17	97.14	97.89
	500	99.58	99.58	99.97	92.51	99.58	76.78	99.61	99.58	99.58	99.63
2.5748	250	97.16	97.18	98.89	81.00	97.16	66.46	97.96	97.16	97.11	98.55
	500	99.58	99.58	99.98	89.59	99.58	75.07	99.63	99.58	99.58	99.77
2.8458	250	97.15	97.18	99.00	77.92	97.16	65.26	98.64	97.15	97.06	94.08
	500	99.58	99.58	99.99	86.56	99.58	73.50	99.70	99.58	99.57	99.42
3.1168	250	97.14	97.18	99.06	75.25	97.15	64.24	16.90	97.14	96.91	95.92
	500	99.58	99.58	99.99	83.61	99.58	72.08	60.47	99.58	99.57	99.51

Table 4-2 demonstrates

- Convergence of all adaptive algorithms is similar except for the RQ and IT algorithms.

- Other than IT, all algorithms converge with a time constant $\propto 1/\lambda_1$.

- For the IT algorithm, the time constant of the principal eigenvector is 1. Since $\lambda_1=17.9$, the convergence of all algorithms is faster than IT.

- For the RQ and LUO algorithms, the time constants are $\|\mathbf{w}_0\|^2/\lambda_1$ and $\|\mathbf{w}_0\|^{-2}/\lambda_1$, respectively.

- Clearly, for larger $\|\mathbf{w}_0\|$, RQ converges at a slower rate than other algorithms and LUO converges faster than other algorithms whose time constants are $1/\lambda_1$.

- For very large $\|\mathbf{w}_0\|$ such as $\|\mathbf{w}_0\|=10.0$, the LUO algorithm fails to converge for $\eta_k=1/(200+k)$ because the convergence becomes unstable.

- For smaller $\|\mathbf{w}_0\|$, the convergence of RQ is better than other algorithms since its time constant $\|\mathbf{w}_0\|^2/\lambda_1$ is smaller than other algorithms whose time constants are $1/\lambda_1$.

Experiments with Various Data Sets: Set 1

[Chatterjee, *Neural Networks*, Vol. 18, No. 2, pp. 145-149, March 2005].

Here I used the covariance matrix (4.20) and added a symmetric matrix $c*R$, where R is a uniform $(0,1)$ random symmetric matrix and $c\in[0.05,2.0]$. I generated 12 sets of 1,000 samples of 10-dimensional Gaussian data with mean zero and random covariance matrix described in (4.20). I chose the starting vector $\mathbf{w}_0 = 0.5*\mathbf{r}$, where $\mathbf{r}\in\Re^{10}$ is a $N(0,1)$ random vector. I used $\eta_k=1/(200+k)$, and for the PF, AL1, and AL2 algorithms, I chose $\mu=10$. I generated the percentage direction cosines (4.21) for all algorithms on each data set and reported the results in Table 4-3. For each data set, I stated the largest 2 eigenvalues λ_1 and λ_2.

Table 4-3. *Convergence of the Principal Eigenvector of A by Adaptive Algorithms at Sample Values* $k=\{250,500\}$ *for Different Data Sets*

λ_1, λ_2	k	OJA	OJAN	LUO	RQ	OJA+	IT	XU	PF	AL1	AL2
11.58, 6.32	250	95.66	95.67	97.61	91.70	95.66	72.19	95.97	95.66	95.65	95.82
	500	99.50	99.50	99.86	98.14	99.50	80.97	99.51	99.50	99.50	99.51
11.63, 6.49	250	95.50	95.54	96.93	91.65	95.51	70.57	96.34	95.51	95.46	96.18
	500	99.51	99.51	99.87	98.28	99.51	80.39	99.54	99.51	99.51	99.54
11.73, 6.92	250	87.62	86.61	97.10	56.91	87.54	47.80	46.00	86.78	88.08	36.39
	500	98.94	98.89	99.80	91.38	98.93	60.20	96.82	98.90	98.96	98.06
11.84, 7.18	250	96.72	96.71	96.53	95.83	96.72	73.84	96.54	96.72	96.73	96.60
	500	99.30	99.30	99.75	98.69	99.30	82.76	99.28	99.30	99.30	99.29
12.14, 7.64	250	96.23	96.22	95.66	95.62	96.23	71.62	96.01	96.22	96.24	96.06
	500	99.10	99.10	99.67	98.54	99.10	81.39	99.08	99.10	99.10	99.08
12.52, 8.08	250	95.10	95.13	95.63	94.45	95.10	60.71	95.78	95.10	95.06	95.67
	500	98.91	98.91	99.62	99.23	98.91	73.59	98.96	98.91	98.91	98.95
12.87, 8.67	250	95.38	95.37	93.65	96.51	95.38	68.84	95.08	95.37	95.39	95.14
	500	98.57	98.57	99.39	98.60	98.57	79.91	98.53	98.57	98.57	98.54

13.57, 9.33	250	94.83	94.82	92.88	97.00	94.82	64.05	94.56	94.82	94.84	94.60
	500	98.35	98.35	99.30	98.66	98.35	76.67	98.32	98.35	98.35	98.32
14.09, 9.88	250	95.95	95.97	92.21	94.23	95.94	40.40	96.05	95.99	95.97	95.86
	500	98.33	98.33	99.17	99.82	98.33	50.26	98.34	98.33	98.33	98.31
17.97, 11.66	250	67.18	72.82	94.87	58.69	67.39	2.78	88.60	74.24	66.40	86.49
	500	98.20	98.45	99.83	90.90	98.21	1.22	99.05	98.50	98.16	98.97
21.54, 12.72	250	95.14	95.19	97.25	84.30	95.14	0.35	95.66	95.19	95.12	95.57
	500	99.79	99.79	99.96	98.49	99.79	3.94	99.80	99.79	99.79	99.79
25.66, 12.92	250	98.23	98.26	99.04	93.90	98.23	2.53	98.42	98.26	98.23	98.37
	500	99.96	99.96	99.99	99.79	99.96	7.62	99.96	99.96	99.96	99.96

Once again, we observe that all algorithms converge in a similar manner except for the RQ and IT algorithms. Out of these two, RQ converges much better than IT, where IT fails to converge for some data sets. Of the remaining algorithms, LUO converges better than the rest for all data sets.

Experiments with Various Data Sets: Set 2

[Chatterjee, *Neural Networks*, Vol. 18, No. 2, pp. 145-149, March 2005].

I further generated 12 sets of 500 samples of 10-dimensional Gaussian data with a mean zero and random covariance matrix (4.20). Here I computed the eigenvectors and eigenvalues of the covariance matrix (4.20). Next, I changed the first two principal eigenvalues of (4.20) to $\lambda_1 = 25$ and $\lambda_2 = \lambda_1/c$, where $c \in [1.1, 10.0]$, and generated the data sets with the new eigenvalues and eigenvectors computed before. For all adaptive algorithms I used $\mathbf{w}_0$, η_k, and μ as described in Section 4.11.2. See Table 4-4.

Observe the following:

Table 4-4. *Convergence of the Principal Eigenvector of A by Adaptive Algorithms at Sample Values k={50,100} for Different Data Sets with Varying λ_1/λ_2*

λ_1/λ_2	k	OJA	OJAN	LUO	RQ	OJA+	IT	XU	PF	AL1	AL2
1.1	50	87.75	87.64	94.14	67.29	87.75	7.60	85.95	87.52	87.78	86.37
	100	96.65	96.64	97.40	90.38	96.65	5.57	96.53	96.64	96.65	96.55
1.5	50	86.07	86.06	91.10	75.70	86.08	39.67	86.29	85.99	86.03	86.37
	100	96.29	96.29	97.53	92.37	96.29	47.73	96.30	96.28	96.28	96.31
2.0	50	92.43	92.39	94.56	83.52	92.43	48.30	91.99	92.34	92.42	92.19
	100	98.04	98.04	98.60	96.00	98.04	55.39	98.00	98.04	98.04	98.01
2.5	50	93.28	93.24	95.03	86.76	93.29	55.77	93.02	93.16	93.23	93.53
	100	98.49	98.49	98.95	96.85	98.49	62.59	98.49	98.49	98.49	98.50
3.0	50	94.39	94.37	96.02	89.18	94.39	60.94	94.50	94.34	94.36	94.53
	100	98.78	98.78	99.08	97.69	98.78	67.92	98.77	98.78	98.78	98.78
4.0	50	96.00	95.99	96.66	90.91	96.01	62.67	95.87	95.96	95.99	96.03
	100	98.96	98.96	99.15	98.40	98.96	69.62	98.97	98.96	98.96	98.96

(continued)

Table 4-4. (*continued*)

5.0	50	94.55	94.55	96.64	89.99	94.55	65.33	94.89	94.52	94.52	94.84
	100	98.93	98.93	99.16	97.85	98.93	71.30	98.94	98.93	98.93	98.93
6.0	50	98.73	98.74	97.62	96.32	98.73	65.37	98.75	98.76	98.74	98.69
	100	99.19	99.19	99.26	99.38	99.19	72.96	99.19	99.19	99.19	99.19
7.0	50	99.36	99.37	98.00	96.29	99.36	64.25	99.41	99.39	99.37	99.33
	100	99.26	99.26	99.29	99.71	99.26	73.32	99.27	99.26	99.26	99.26
8.0	50	97.12	97.11	97.97	92.96	97.12	62.88	97.12	97.09	97.11	97.17
	100	99.25	99.25	99.34	98.71	99.25	70.05	99.25	99.25	99.25	99.25
9.0	50	97.32	97.31	98.18	92.85	97.32	63.17	97.27	97.28	97.31	97.34
	100	99.33	99.33	99.41	98.87	99.33	70.33	99.33	99.33	99.33	99.33
10.0	50	97.82	97.81	98.43	94.38	97.82	66.24	97.70	97.79	97.81	97.79
	100	99.43	99.43	99.49	99.04	99.43	73.12	99.43	99.43	99.43	99.43

- The convergences are similar to dataset 1.

- The convergence improves for larger values of k and for larger ratios of λ_1/λ_2 as supported by Table 4-1 and the experimental results in Table 4-4.

Experiments with Real-World Non-Stationary Data

In these experiments I use real-world non-stationary data from Publicly Real-World Datasets to Evaluate Stream Learning Algorithms [Vinicius Souza *et al.* 2020], INSECTS-incremental-abrupt_balanced_norm.arff. It is important to demonstrate the performance of these algorithms on real non-stationary data since in practical edge applications the data is usually time varying and changes over time.

The data has 33 components and 80,000 samples. It contains periodic abrupt changes. Figure 4-3 shows the components.

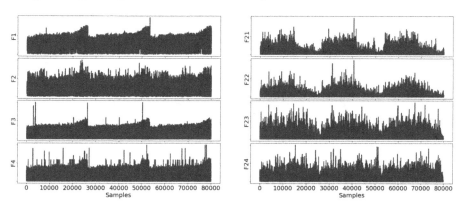

Figure 4-3. *Non-stationary real-world data with abrupt periodic changes*

The first eight eigenvalues of the correlation matrix of all samples are

[18.704, 10.473, 8.994, 7.862, 7.276 6.636 5.565 4.894, ...].

I used the adaptive algorithms discussed in this chapter and plotted the percentage direction cosine (4.21). Figure 4-4 shows that all algorithms converged well in spite of the non-stationarity in the data.

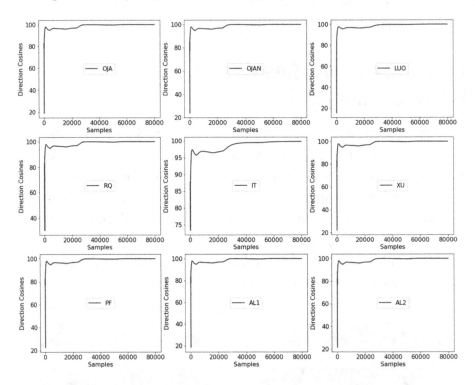

Figure 4-4. *Convergence of the adaptive algorithms for first principal eigenvector on real-world non-stationary data (ideal value=100)*

4.12 Concluding Remarks

The RQ, OJAN, and LUO algorithms require $\|\mathbf{w}_0\|=1$ for the principal eigenvector $\mathbf{w}_k$ to converge to unit norm (i.e., $\|\mathbf{w}_k\| \to 1$). All other algorithms (except PF) converged to the unit norm for various non-zero values for $\mathbf{w}_0$. As per Theorem 4.8, the PF algorithm successfully converged to $\|\mathbf{w}_k\| \to \sqrt{1 + \lambda_1 / \mu}$.

The unified framework and the availability of the objective functions allows me to derive and analyze each algorithm on its convergence performance. The results for each algorithm are summarized in Table 4-5. For each method, I computed the approximate MATLAB flop[2] count for one iteration of the adaptive algorithm for $n=10$ and show them in Table 4-5.

Table 4-5. *Comparison of Different Adaptive Algorithms [Chatterjee, Neural Networks 2005]*

Alg.	Pros	Cons
OJA	Convergence increases with larger λ_1 and λ_1/λ_2. Upper bound of η_k can be determined. Fewer computations per iteration (Flops=460).	Convergence cannot be improved by larger $\|\mathbf{w}_0\|$.
OJAN	Convergence increases with larger λ_1 and λ_1/λ_2. Fewer computations per iteration (Flops=481).	Upper bound of η_k not available. Convergence cannot be improved by larger $\|\mathbf{w}_0\|$. Require $\|\mathbf{w}_0\|=1$ for $\mathbf{w}_k$ to converge to unit norm.
LUO	Convergence increases with larger λ_1 and λ_1/λ_2. Convergence increases for larger $\|\mathbf{w}_0\|$. Fewer computations per iteration (Flops=502).	Upper bound of η_k not available. Convergence decreases for smaller $\|\mathbf{w}_0\|$. Requires $\|\mathbf{w}_0\|=1$ for $\mathbf{w}_k$ to converge to unit norm.

(continued)

---

[2] Flop is a floating-point operation. Addition, subtraction, multiplication, and division of real numbers are one flop each.

Table 4-5. (*continued*)

Alg.	Pros	Cons
RQ	Convergence increases with larger λ_1 and λ_1/λ_2. Convergence increases for smaller $\|\mathbf{w}_0\|$. Fewer computations per iteration (Flops=503).	Upper bound of η_k not available. Convergence decreases for larger v. Requires $\|\mathbf{w}_0\|=1$ for $\mathbf{w}_k$ to converge to unit norm.
OJA+	Convergence increases with larger λ_1 and λ_1/λ_2. Upper bound of η_k can be determined. Fewer computations per iteration (Flops=501).	Convergence cannot be improved by larger $\|\mathbf{w}_0\|$.
IT	Upper bound of η_k can be determined. Fewer computations per iteration (Flops=460).	Convergence independent of λ_1. Experimental results show poor convergence.
XU	Convergence increases with larger λ_1 and λ_1/λ_2. Upper bound of η_k can be determined.	Convergence cannot be improved by larger $\|\mathbf{w}_0\|$. More computations per iteration (Flops=800).

(*continued*)

Table 4-5. (*continued*)

Alg.	Pros	Cons
PF	Convergence increases with larger λ_1 and λ_1/λ_2. Convergence increases with μ. Upper bound of η_k can be determined. Smallest computations per iteration (Flops=271).	Convergence cannot be improved by larger $\|\mathbf{w}_0\|$. $\mathbf{w}_k$ does not converge to unit norm.
AL1	Convergence increases with larger λ_1 and λ_1/λ_2. Convergence increases with μ. Upper bound of η_k can be determined. Fewer computations per iteration (Flops=511).	Convergence cannot be improved by larger $\|\mathbf{w}_0\|$.
AL2	Convergence increases with larger λ_1 and λ_1/λ_2. Convergence increases with μ. Upper bound of η_k can be determined.	Convergence cannot be improved by larger $\|\mathbf{w}_0\|$. Largest computations per iteration (Flops=851).

In summary, I discussed ten adaptive algorithms for PCA, some of them new, from a common framework with an objective function for each. Note that although I applied the gradient descent technique on these objective functions, I could have applied any other technique of nonlinear optimization such as steepest descent, conjugate direction, Newton-Raphson, or recursive least squares.

CHAPTER 5

Principal and Minor Eigenvectors

5.1 Introduction and Use Cases

In Chapter 4, I discussed adaptive algorithms for the computation of the principal eigenvector of the online correlation matrix $A_k \in \Re^{nXn}$. However, in some applications, it is not enough to just compute the principal eigenvector; we also need to compute the minor eigenvectors of A_k. One such application is multi-dimensional data compression or data dimensionality reduction in multimedia video transmission [Le Gall 91]. For example, in still video compression by the JPEG technique, the image is divided into 8X8 blocks. This high-dimensional video data can be reduced to lower dimensions by projecting it onto the principal eigenvector subspace of its online correlation matrix. The process of data projection onto the eigenvector subspace by a linear transform is known as principal component analysis (PCA) and is closely related to the Karhunen-Loeve Transform (KLT) [Fukunaga 90]. We approximate the eigenvectors by fixed transform vectors given by the discrete cosine transform (DCT), which is the central compression method of the MPEG standard [Le Gall 91]. It can be shown that DCT is asymptotically equivalent to PCA for signals coming from a first-order Markov model, which is a reasonable model for digital images.

© Chanchal Chatterjee 2022
C. Chatterjee, *Adaptive Machine Learning Algorithms with Python*,
https://doi.org/10.1007/978-1-4842-8017-1_5

Figure 5-1 shows the original 128-dimensional signal on the left. On the right, using adaptive algorithms, I reconstructed the signal with just 16 principal components with an 8x compression. Results show that the reconstructed data is near identical.

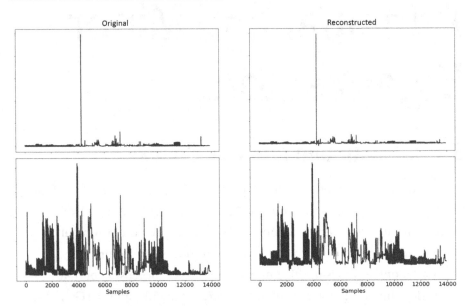

Figure 5-1. *Original signal on the left and reconstructed signal on the right after 8x compression with principal components*

More details on this application and code are given in Chapter 8.

Figure 5-2 shows the number 9 from the Keras MNIST dataset [Keras, MNIST]. The original data is 28x28=784 dimensional. I used the first 100 principal components and reconstructed the images. This is 7.8x reduction in data. The reconstructed images look similar to the original.

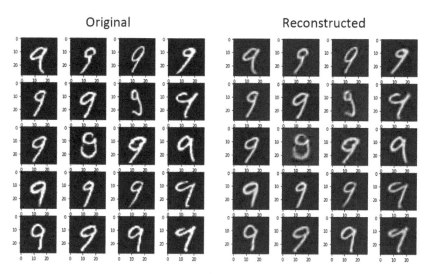

Figure 5-2. *Original MNIST number 9 on the left and reconstructed number on the right after 7.8x compression with principal components*

The following Python code can be used to PCA compress the data X[nDim,nSamples]:

```
# Compute the first 100 principal components and PCA compress
the data from scipy.linalg import eigh
nSamples = X.shape[1]
nDim = X.shape[0]
corX = (X @ X.T) / nSamples
eigvals, eigvecs = eigh(corX)
V  = np.fliplr(eigvecs)
# PCA Transformed data
Y = V[:,:100] @ V[:,:100].T @ X
```

Since PCA is optimal in the mean-squares error sense, it has been widely used in signal and image processing, data analysis, pattern recognition, communications, and control engineering. Applications such as the separation of signal subspace and noise subspace are important in digital communications and digital image processing. Applications of PCA in signal processing include temporal and spatial domain spectral analyses. Examples include multiple signal classification (MUSIC) techniques, minimum-norm methods, ESPIRIT estimators, and weighted subspace fitting (WSF) methods for estimating frequencies of sinusoids or direction of arrival (DOA) of plane waves impinging on an antenna array. More recently, PCA is applied to blind digital communications and speech processing. For example, in [Chatterjee et al. 97-99; Chen et al. 99], PCA is employed in a blind two-dimensional RAKE receiver for DS-CDMA-based space-frequency processing to solve the near-far problem. In [Diamantaras and Strintzis 97], PCA is used for optimal linear vector coding and decoding in the presence of noise.

Unified Framework

In this chapter, I present a unified framework to derive and analyze several algorithms (some well-known) for adaptive eigen-decomposition. The steps consist of the following:

1. Description of an objective function from which I derive the adaptive algorithm by using the gradient descent method of nonlinear optimization.

2. For each objective function, I offer three methods of deriving the adaptive algorithms, which lead to three sets of adaptive algorithms, each with its own computational cost, convergence property, and implementation simplicity:

2.1. *Homogeneous Adaptive Rule:* These algorithms do not compute the true normalized eigenvectors with decreasing eigenvalues. Instead, they produce a linear combination of the unit eigenvectors. However, these algorithms can be computed quickly and implemented in parallel networks.

2.2. *Deflation Adaptive Rule:* Here, we lose the homogeneity of the algorithms, but produce unit eigenvectors with decreasing eigenvalues, thereby satisfying many application requirements. However, the training is sequential, and the learning rule for the weights of the nth neuron depends on the training of the previous $n-1$ neurons, thereby making the training process difficult for parallel implementations.

2.3. *Weighted Adaptive Rule:* These algorithms are obtained by breaking the symmetry of the homogeneous algorithms by using a different scalar weight for each eigenvector. The unit eigenvectors are obtained explicitly and in the order of decreasing eigenvalues. It is also possible to implement these algorithms in parallel networks. However, the algorithms require extra computation.

It is interesting to observe that many algorithms presented here were independent discoveries over more than a decade. For example, Oja et al. [Oja and Karhunen 85] first presented the Oja homogeneous algorithm (5.5) in 1985. Various practitioners extensively analyzed this algorithm in separate publications from 1989 to 1998 (see Section 5.3.1). Sanger [Sanger 89] showed the deflation version (5.8) of this algorithm in 1989. Subsequently, Oja, Brockett, Xu and Chen et al. presented the weighted variation (5.10) of this algorithm during 1992-98 (see Sec. 5.3.2). This chapter unifies these algorithms and many others, including several new algorithms, in a single framework with uncomplicated derivations. Note that although these algorithms were discovered from various perspectives, I presented them on a single foundation.

Given an asymptotically stationary sequence $\{\mathbf{x}_k \in \Re^n\}$ that has been centered to zero mean, the asymptotic data correlation matrix is given by $A = \lim_{k \to \infty} E\left[\mathbf{x}_k \mathbf{x}_k^T \right]$. The $p \leq n$ orthogonal unit principal eigenvectors $\boldsymbol{\phi}_1, \ldots, \boldsymbol{\phi}_p$ of A are given by

$$A\boldsymbol{\phi}_i = \lambda_i \boldsymbol{\phi}_i, \ \boldsymbol{\phi}_i^T A\boldsymbol{\phi}_j = \lambda_i \delta_{ij}, \text{ and } \boldsymbol{\phi}_i^T \boldsymbol{\phi}_j = \delta_{ij} \text{ for } i=1,\ldots,p, \tag{5.1}$$

where $\lambda_1 > \ldots > \lambda_p > \lambda_{p+1} \geq \ldots \geq \lambda_n > 0$ are the p largest eigenvalues of A in descending order of magnitude. If the sequence $\{\mathbf{x}_k\}$ is non-stationary, we compute the online data correlation matrix $A_k \in \Re^{n \times n}$ by (2.3) or (2.5).

In my analyses of the algorithms, I follow the methodology outlined in Section 1.4. For the algorithms, I describe an *objective function* $J(\mathbf{w}_i; A)$ and an updated rule of the form

$$W_{k+1} = W_k + \eta_k h(W_k, A_k), \tag{5.2}$$

where $h(W_k, A_k)$ follows certain continuity and regularity properties [Ljung 77, 92] and η_k is a decreasing gain sequence.

Outline of This Chapter

In Section 5.2, I discuss the objective functions and their variations that are analyzed in this chapter. In Section 5.3, I present adaptive algorithms for the homogeneous, deflation, and weighted variations for the OJA objective function with convergence results. In Section 5.4, I analyze the same three variations for the mean squared error (XU) objective function. In Section 5.5, I discuss algorithms derived from the penalty function (PF) objective function and prove the convergence results. In Section 5.6, I consider the augmented Lagrangian 1 (AL1) objective function; in Section 5.7, I present the augmented Lagrangian 2 (AL2) objective function. In Section 5.8, I present the information theory (IT) criterion; in Section 5.9, I describe the Rayleigh quotient (RQ) criterion. In Section 5.10, I present a summary of all algorithms discussed here. In Section 5.11, I discuss the experimental results and I conclude the chapter with Section 5.12.

5.2 Algorithms and Objective Functions

Conforming to my proposed methodology in Chapter 2.2, for each algorithm, I describe an objective function and derive the adaptive algorithm for it. I have itemized the algorithms based on their inventors or on the objective functions from which they are derived. The objective functions are

1. Oja's objective function [Oja 92; Oja et al. 92],

2. Mean squared error objective function for Xu's Algorithm [Xu 91, 93],

3. Penalty function method [Chauvin 89; Mathew et al. 95],

4. Augmented Lagrangian Method 1,

5. Augmented Lagrangian Method 2,

6. Information theory criterion [Pumbley 95; Miao and Hua 98],

7. Rayleigh quotient criterion [Luo 97; Sarkar and Yang 89; Yang et al. 89; Fu and Dowling 94, 95].

Summary of Objective Functions for Adaptive Algorithms

Each algorithm is of the form (5.2), which we derive from an objective function $J(\mathbf{w}_i; A)$, where $W = [\mathbf{w}_1, ..., \mathbf{w}_p]$ ($p \leq n$). The objective function $J(\mathbf{w}_i; A)$ for each adaptive algorithm is given in Table 5-1.

Table 5-1. *Objective Functions for Adaptive Eigen-Decomposition Algorithms Discussed Here*

Alg.	Type	Objective Function $J(\cdot)$
OJA	Homogeneous	$-\mathbf{w}_i^T A^2 \mathbf{w}_i + \dfrac{1}{2}\left(\mathbf{w}_i^T A \mathbf{w}_i\right)^2 + \displaystyle\sum_{j=1, j\neq i}^{p} \left(\mathbf{w}_i^T A \mathbf{w}_j\right)^2$
	Deflation	$-\mathbf{w}_i^T A^2 \mathbf{w}_i + \dfrac{1}{2}\left(\mathbf{w}_i^T A \mathbf{w}_i\right)^2 + \displaystyle\sum_{j=1}^{i-1}\left(\mathbf{w}_i^T A \mathbf{w}_j\right)^2$
	Weighted	$-c_i\mathbf{w}_i^T A^2 \mathbf{w}_i + \dfrac{c_i}{2}\left(\mathbf{w}_i^T A \mathbf{w}_i\right)^2 + \displaystyle\sum_{j=1, j\neq i}^{p} c_j\left(\mathbf{w}_i^T A \mathbf{w}_j\right)^2$
XU	Homogeneous	$-\mathbf{w}_i^T A \mathbf{w}_i + \left(\mathbf{w}_i^T A \mathbf{w}_i\right)\left(\mathbf{w}_i^T \mathbf{w}_i - 1\right) + 2\displaystyle\sum_{j=1, j\neq i}^{p} \mathbf{w}_i^T A \mathbf{w}_j \mathbf{w}_j^T \mathbf{w}_i$
	Deflation	$-\mathbf{w}_i^T A \mathbf{w}_i + \left(\mathbf{w}_i^T A \mathbf{w}_i\right)\left(\mathbf{w}_i^T \mathbf{w}_i - 1\right) + 2\displaystyle\sum_{j=1}^{i-1} \mathbf{w}_i^T A \mathbf{w}_j \mathbf{w}_j^T \mathbf{w}_i$
	Weighted	$-c_i\mathbf{w}_i^T A \mathbf{w}_i + c_i\left(\mathbf{w}_i^T A \mathbf{w}_i\right)\left(\mathbf{w}_i^T \mathbf{w}_i - 1\right)$ $+ 2\displaystyle\sum_{j=1, j\neq i}^{p} c_j \mathbf{w}_i^T A \mathbf{w}_j \mathbf{w}_j^T \mathbf{w}_i$
PF	Homogeneous	$-\mathbf{w}_i^T A \mathbf{w}_i + \mu \Lambda_H\left(\mathbf{w}_1, \ldots, \mathbf{w}_p\right)$
	Deflation	$-\mathbf{w}_i^T A \mathbf{w}_i + \mu \Lambda_D\left(\mathbf{w}_1, \ldots, \mathbf{w}_p\right)$
	Weighted	$-c_i\mathbf{w}_i^T A \mathbf{w}_i + \mu \Lambda_W\left(\mathbf{w}_1, \ldots, \mathbf{w}_p\right)$

(continued)

Table 5-1. (*continued*)

Alg.	Type	Objective Function $J(\cdot)$
AL1	Homogeneous	$-\mathbf{w}_i^T A \mathbf{w}_i + \alpha \left(\mathbf{w}_i^T \mathbf{w}_i - 1 \right) + 2 \sum\limits_{j=1, j \neq i}^{p} \beta_j \mathbf{w}_j^T \mathbf{w}_i$ $+ \mu \Lambda_H \left(\mathbf{w}_1, \ldots, \mathbf{w}_p \right)$
	Deflation	$-\mathbf{w}_i^T A \mathbf{w}_i + \alpha \left(\mathbf{w}_i^T \mathbf{w}_i - 1 \right) + 2 \sum\limits_{j=1}^{i-1} \beta_j \mathbf{w}_j^T \mathbf{w}_i$ $+ \mu \Lambda_D \left(\mathbf{w}_1, \ldots, \mathbf{w}_i \right)$
	Weighted	$-c_i \mathbf{w}_i^T A \mathbf{w}_i + \alpha c_i \left(\mathbf{w}_i^T \mathbf{w}_i - 1 \right) + 2 \sum\limits_{j=1, j \neq i}^{p} \beta_j c_j \mathbf{w}_j^T \mathbf{w}_i$ $+ \mu \Lambda_W \left(\mathbf{w}_1, \ldots, \mathbf{w}_p \right)$
AL2	Homogeneous	$-\mathbf{w}_i^T A \mathbf{w}_i + \left(\mathbf{w}_i^T A \mathbf{w}_i \right)_i \left(\mathbf{w}_i^T \mathbf{w}_i - 1 \right) + 2 \sum\limits_{j=1, j \neq i}^{p} \mathbf{w}_i^T A \mathbf{w}_j \mathbf{w}_j^T \mathbf{w}_i +$ $+ \mu \Lambda_H \left(\mathbf{w}_1, \ldots, \mathbf{w}_p \right)$
	Deflation	$-\mathbf{w}_i^T A \mathbf{w}_k^i + \left(\mathbf{w}_i^T A \mathbf{w}_i \right) \left(\mathbf{w}_i^T \mathbf{w}_i - 1 \right) + 2 \sum\limits_{j=1}^{i-1} \mathbf{w}_i^T A \mathbf{w}_j \mathbf{w}_j^T \mathbf{w}_i +$ $+ \mu \Lambda_D \left(\mathbf{w}_1, \ldots, \mathbf{w}_i \right)$
	Weighted	$-c_i \mathbf{w}_i^T A \mathbf{w}_i + c_i \left(\mathbf{w}_i^T A \mathbf{w}_i \right) \left(\mathbf{w}_i^T \mathbf{w}_i - 1 \right)$ $+ 2 \sum\limits_{j=1, j \neq i}^{p} c_j \mathbf{w}_i^T A \mathbf{w}_j \mathbf{w}_j^T \mathbf{w}_i + \mu \Lambda_w \left(\mathbf{w}_1, \ldots \mathbf{w}_p \right)$

(*continued*)

Table 5-1. (*continued*)

Alg.	Type	Objective Function $J(\cdot)$
IT	Homogeneous	$\mathbf{w}_i^T\mathbf{w}_i - \log\left(\mathbf{w}_i^T A \mathbf{w}_i\right) + \alpha\left(\mathbf{w}_i^T\mathbf{w}_i - 1\right) + 2\sum_{j=1,j\neq i}^{p}\beta_j\mathbf{w}_j^T\mathbf{w}_i$
	Deflation	$\mathbf{w}_i^T\mathbf{w}_i - \log\left(\mathbf{w}_i^T A \mathbf{w}_i\right) + \alpha\left(\mathbf{w}_i^T\mathbf{w}_i - 1\right) + 2\sum_{j=1}^{i-1}\beta_j\mathbf{w}_j^T\mathbf{w}_i$
	Weighted	$c_i\mathbf{w}_i^T\mathbf{w}_i - c_i\log\left(\mathbf{w}_i^T A \mathbf{w}_i\right) + \alpha c_i\left(\mathbf{w}_i^T\mathbf{w}_i - 1\right)$ $+ 2\sum_{j=1,j\neq i}^{p}\beta_j c_j\mathbf{w}_j^T\mathbf{w}_i$
RQ	Homogeneous	$-\left(\mathbf{w}_i^T A \mathbf{w}_i / \mathbf{w}_i^T\mathbf{w}_i\right) + \alpha\left(\mathbf{w}_i^T\mathbf{w}_i - 1\right) + 2\sum_{j=1,j\neq i}^{p}\beta_j\mathbf{w}_j^T\mathbf{w}_i$
	Deflation	$-\left(\mathbf{w}_i^T A \mathbf{w}_i / \mathbf{w}_i^T\mathbf{w}_i\right) + \alpha\left(\mathbf{w}_i^T\mathbf{w}_i - 1\right) + 2\sum_{j=1}^{i-1}\beta_j\mathbf{w}_j^T\mathbf{w}_i$
	Weighted	$-c_i\left(\mathbf{w}_i^T A \mathbf{w}_i / \mathbf{w}_i^T\mathbf{w}_i\right) + \alpha c_i\left(\mathbf{w}_i^T\mathbf{w}_i - 1\right) + 2\sum_{j=1,j\neq i}^{p}\beta_j c_j\mathbf{w}_j^T\mathbf{w}_i$

In these expressions, Λ_H, Λ_D, Λ_W are defined as

$$\Lambda_H\left(\mathbf{w}_1,\ldots,\mathbf{w}_p\right) = \sum_{j=1,j\neq i}^{p}\left(\mathbf{w}_j^T\mathbf{w}_i\right)^2 + \frac{1}{2}\left(\mathbf{w}_i^T\mathbf{w}_i - 1\right)^2,$$

$$\Lambda_D\left(\mathbf{w}_1,\ldots,\mathbf{w}_i\right) = \sum_{j=1}^{i-1}\left(\mathbf{w}_j^T\mathbf{w}_i\right)^2 + \frac{1}{2}\left(\mathbf{w}_i^T\mathbf{w}_i - 1\right)^2,$$

and

$$\Lambda_W\left(\mathbf{w}_1,\ldots,\mathbf{w}_p\right) = \sum_{j=1,j\neq i}^{p}c_j\left(\mathbf{w}_j^T\mathbf{w}_i\right)^2 + \frac{c_i}{2}\left(\mathbf{w}_i^T\mathbf{w}_i - 1\right)^2.$$

5.3 OJA Algorithms

OJA Homogeneous Algorithm

The objective function for the OJA homogeneous algorithm is

$$J\left(\mathbf{w}_k^i; A_k\right) = -{\mathbf{w}_k^i}^T A_k^2 \mathbf{w}_k^i + \frac{1}{2}\left({\mathbf{w}_k^i}^T A_k \mathbf{w}_k^i\right)^2 + \sum_{j=1, j\neq i}^{p}\left({\mathbf{w}_k^i}^T A_k \mathbf{w}_k^j\right)^2, \qquad (5.3)$$

for $i=1,...,p$ $(p\leq n)$. From the gradient of (5.3) with respect to $\mathbf{w}_k^i$ we obtain the following adaptive algorithms:

$$\mathbf{w}_{k+1}^i = \mathbf{w}_k^i - \eta_k A_k^{-1}\nabla_{\mathbf{w}_k^i} J\left(\mathbf{w}_k^i; A_k\right) \text{ for } i=1,...,p,$$

or

$$\mathbf{w}_{k+1}^i = \mathbf{w}_k^i + \eta_k\left(A_k \mathbf{w}_k^i - \sum_{j=1}^{p}\mathbf{w}_k^j {\mathbf{w}_k^j}^T A_k \mathbf{w}_k^i\right) \qquad (5.4)$$

for $i=1,...,p$, where η_k is a small decreasing constant. We define a matrix $W_k = \left[\mathbf{w}_k^1...\mathbf{w}_k^p\right]$ $(p\leq n)$, for which the columns are the p weight vectors that converge to the p principal eigenvectors of A respectively. We can represent (5.4) as

$$W_{k+1} = W_k + \eta_k\left(A_k W_k - W_k W_k^T A_k W_k\right). \qquad (5.5)$$

This is the matrix form of the *principal subspace learning algorithm* given by Oja [Oja 85,89,92].

W_k converges $W^* = \Phi D U$, where $D=[D_1|0]^T\in\mathfrak{R}^{n\times p}$, $D_1=diag(d_1,...,d_p)\in\mathfrak{R}^{p\times p}$, $d_i=\pm 1$ for $i=1,...,p$, and $U\in\mathfrak{R}^{p\times p}$ is an arbitrary rotation matrix (i.e., $U^T U = UU^T = I_p$).

OJA Deflation Algorithm

The objective function for the OJA deflation adaptive PCA algorithm is

$$J\left(\mathbf{w}_k^i; A_k\right) = -\mathbf{w}_k^{i^T} A_k^2 \mathbf{w}_k^i + \frac{1}{2}\left(\mathbf{w}_k^{i^T} A_k \mathbf{w}_k^i\right)^2 + \sum_{j=1}^{i-1}\left(\mathbf{w}_k^{i^T} A_k \mathbf{w}_k^j\right)^2 \quad \text{for } i=1,\dots,p. \quad (5.6)$$

From the gradient of (5.6), we obtain the OJA deflation adaptive gradient descent algorithm:

$$W_{k+1} = W_k + \eta_k\left(A_k W_k - W_k \mathrm{UT}\left(W_k^T A_k W_k\right)\right), \quad (5.7)$$

where UT[·] sets all elements below the diagonal of its matrix argument to zero, thereby making it upper triangular. This algorithm is also known as the *generalized Hebbian algorithm* [Sanger 89]. Sanger proved that W_k converges to $W^* = [\pm\boldsymbol{\phi}_1 \pm\boldsymbol{\phi}_2 \dots \pm\boldsymbol{\phi}_p]$ as $k\to\infty$.

OJA Weighted Algorithm

The objective function for the OJA weighted adaptive PCA algorithm is

$$J\left(\mathbf{w}_k^i; A_k\right) = -c_i\mathbf{w}_k^{i^T} A_k^2 \mathbf{w}_k^i + \frac{c_i}{2}\left(\mathbf{w}_k^{i^T} A_k \mathbf{w}_k^i\right)^2 + \sum_{j=1, j\neq i}^{p} c_j\left(\mathbf{w}_k^{i^T} A_k \mathbf{w}_k^j\right)^2, \quad (5.8)$$

for $i=1,\dots,p$ and $c_1,\dots,c_p$ are small positive numbers satisfying $c_1>c_2>\dots>c_p>0$, $p\leq n$. From (5.8), we obtain the OJA weighted adaptive gradient descent algorithm for PCA as

$$W_{k+1} = W_k + \eta_k\left(A_k W_k C - W_k C W_k^T A_k W_k\right), \quad (5.9)$$

where $C=diag(c_1,\dots,c_p)$. This algorithm is also known as *Brockett's subspace* algorithm [Xu 93, Chen, Amari, Lin 98]. Xu [Xu 93, Theorems 5 and 6] proved that W_k converges to $W^* = [\pm\boldsymbol{\phi}_1 \pm\boldsymbol{\phi}_2 \dots \pm\boldsymbol{\phi}_p]$ as $k\to\infty$.

OJA Algorithm Python Code

The following Python code implements the OJA PCA algorithms with data X[nDim,nSamples]:

```python
from numpy import linalg as la
nEA = 4 # number of PCA components computed
nEpochs  = 2
A  = np.zeros(shape=(nDim,nDim)) # stores adaptive
                                correlation matrix
W1 = 0.1 * np.ones(shape=(nDim,nEA)) # weight vectors of all
                                algorithms
W2 = W1
W3 = W1
c = [2-0.3*k for k in range(nEA)]
C = np.diag(c)
for epoch in range(nEpochs):
    for iter in range(nSamples):
        cnt = nSamples*epoch + iter
        # Update data correlation matrix A with current data
        sample x
        x = X[:,iter]
        x = x.reshape(nDim,1)
        A = A + (1.0/(1 + cnt))*((np.dot(x, x.T)) - A)
        # Homogeneous Gradient Descent
        W1 = W1 + (1/(100 + cnt))*(A @ W1 - W1 @ (W1.T
        @ A @ W1))
        # Deflated Gradient Descent
        W2 = W2 + (1/(100 + cnt))*(A @ W2 - W2 @ np.triu(W2.T
        @ A @ W2))
        # Weighted Gradient Descent
        W3 = W3 + (1/(220 + cnt))*(A @ W3 @ C - W3 @ C @ (W3.T
        @ A @ W3))
```

5.4 XU Algorithms

XU Homogeneous Algorithm

The objective function for the XU homogeneous adaptive PCA algorithm is

$$J\left(\mathbf{w}_k^i; A_k\right) = -\mathbf{w}_k^{i^T} A_k \mathbf{w}_k^i + \left(\mathbf{w}_k^{i^T} A_k \mathbf{w}_k^i\right)\left(\mathbf{w}_k^{i^T} \mathbf{w}_k^i - 1\right)$$

$$+ 2 \sum_{j=1, j \neq i}^{p} \mathbf{w}_k^{i^T} A_k \mathbf{w}_k^j \mathbf{w}_k^{j^T} \mathbf{w}_k^i, \tag{5.10}$$

for $i=1,...,p$. From the gradient of (5.10) with respect to $\mathbf{w}_k^i$, we obtain the XU homogeneous adaptive gradient descent algorithm for PCA as

$$W_{k+1} = W_k + \eta_k \left(2 A_k W_k - A_k W_k W_k^T W_k - W_k W_k^T A_k W_k\right). \tag{5.11}$$

This algorithm is also known as the *least mean squared error reconstruction (LMSER) algorithm* [Xu 91,93] and was derived from a least mean squared error criterion of a feed-forward neural network. Xu [Xu 93, Theorems 2, 3] proved that W_k converges to $W^* = \Phi DU$, where $D = [D_1|0]^T \in \mathfrak{R}^{n \times p}$, $D_1 = diag(d_1,...,d_p) \in \mathfrak{R}^{p \times p}$, $d_i = \pm 1$ for $i=1,...,p$, and $U \in \mathfrak{R}^{p \times p}$ is an arbitrary rotation matrix.

XU Deflation Algorithm

The objective function for the XU deflation adaptive PCA algorithm is

$$J\left(\mathbf{w}_k^i; A_k\right) = -\mathbf{w}_k^{i^T} A_k \mathbf{w}_k^i + \left(\mathbf{w}_k^{i^T} A_k \mathbf{w}_k^i\right)\left(\mathbf{w}_k^{i^T} \mathbf{w}_k^i - 1\right) + 2 \sum_{j=1}^{i-1} \mathbf{w}_k^{i^T} A_k \mathbf{w}_k^j \mathbf{w}_k^{j^T} \mathbf{w}_k^i \tag{5.12}$$

for $i=1,...,p$. The XU deflation adaptive gradient descent algorithm for PCA is

$$W_{k+1} = W_k + \eta_k \left(2 A_k W_k - A_k W_k \mathrm{UT}\left(W_k^T W_k\right) - W_k \mathrm{UT}\left(W_k^T A_k W_k\right)\right), \tag{5.13}$$

where UT[·] sets all elements below the diagonal of its matrix argument to zero. Chatterjee et al. [Mar 00, Theorems 1,2] proved that W_k converges to $W^* = [\pm\phi_1 \pm\phi_2 \dots \pm\phi_p]$ as $k\to\infty$.

XU Weighted Algorithm

The objective function for the XU weighted adaptive PCA algorithm is

$$J\left(\mathbf{w}_k^i; A_k\right) = -c_i \mathbf{w}_k^{i\,T} A_k \mathbf{w}_k^i + c_i \left(\mathbf{w}_k^{i\,T} A_k \mathbf{w}_k^i\right)\left(\mathbf{w}_k^{i\,T} \mathbf{w}_k^i - 1\right)$$

$$+ 2 \sum_{j=1, j\neq i}^{p} c_j \mathbf{w}_k^{i\,T} A_k \mathbf{w}_k^j \mathbf{w}_k^{j\,T} \mathbf{w}_k^i \qquad (5.14)$$

for $i=1,\dots,p$ and $c_1 > c_2 > \dots > c_p > 0$. From (5.14), the XU weighted adaptive gradient descent algorithm for PCA is

$$W_{k+1} = W_k + \eta_k \left(2 A_k W_k C_k - W_k C W_k^T A_k W_k - A_k W_k C W_k^T W_k\right), \qquad (5.15)$$

where $C = diag(c_1,\dots,c_p)$. Xu [Xu 93, Theorems 5, 6] proved that W_k converges to $W^* = [\pm\phi_1 \pm\phi_2 \dots \pm\phi_p]$ as $k\to\infty$.

XU Algorithm Python Code

The following Python code implements the XU PCA algorithms with data X[nDim,nSamples]:

```
from numpy import linalg as la
nEA = 4 # number of PCA components computed
nEpochs  = 2
A  = np.zeros(shape=(nDim,nDim)) # stores adaptive
                                 correlation matrix
W1 = 0.1 * np.ones(shape=(nDim,nEA)) # weight vectors of all
                                     algorithms
W2 = W1
W3 = W1
```

```
c = [2-0.3*k for k in range(nEA)]
C = np.diag(c)
for epoch in range(nEpochs):
    for iter in range(nSamples):
        cnt = nSamples*epoch + iter
        x = X[:,iter]
        x = x.reshape(nDim,1)
        A = A + (1.0/(1 + cnt))*((np.dot(x, x.T)) - A)
        # Homogeneous Gradient Descent
        W1 = W1 + (1/(100 + cnt))*(A @ W1 - 0.5 * W1 @ (W1.T
                                  @ A @ W1) -0.5 * A @ W1
                                  @ (W1.T @ W1))
        # Deflated Gradient Descent
        W2 = W2 + (1/(100 + cnt))*(A @ W2 - 0.5 * W2 @
                                  np.triu(W2.T @ A @ W2) -0.5
                                  * A @ W2 @ np.triu(W2.T @ W2))
        # Weighted Gradient Descent
        W3 = W3 + (1/(100 + cnt))*(A @ W3 @ C - 0.5 * W3 @ C
                                  @ (W3.T @ A @ W3) -0.5 * A
                                  @ W3 @ C @ (W3.T @ W3))
```

5.5 PF Algorithms

PF Homogeneous Algorithm

We obtain the objective function for the penalty function homogeneous PCA algorithm by expressing the Rayleigh quotient criterion as the following penalty function:

$$J\left(\mathbf{w}_k^i; A_k\right) = -\mathbf{w}_k^{i^T} A_k \mathbf{w}_k^i + \mu\left(\sum_{j=1,j\neq i}^{p}\left(\mathbf{w}_k^{j^T}\mathbf{w}_k^i\right)^2 + \frac{1}{2}\left(\mathbf{w}_k^{i^T}\mathbf{w}_k^i - 1\right)^2\right), \quad (5.16)$$

where $\mu>0$ and $i=1,...,p$. From the gradient of (5.16) with respect to $\mathbf{w}_k^i$, we obtain the PF homogeneous adaptive gradient descent algorithm for PCA as

$$W_{k+1} = W_k + \eta_k \left(A_k W_k - \mu W_k \left(W_k^T W_k - I_p \right) \right),\tag{5.17}$$

where I_p is a $p \mathrm{X} p$ identity matrix.

W_k converges to $W^* = \Phi D U$, where $D = [D_1|0]^T \in \Re^{n \mathrm{X} p}$, $D_1 = \mathrm{diag}(d_1,...,d_p)$ $\in \Re^{p \mathrm{X} p}$, $d_i = \pm\sqrt{1 + (\lambda_i / \mu)}$ for $i=1,...,p$, and $U \in \Re^{p \mathrm{X} p}$ is an arbitrary rotation matrix. Recall that $\lambda_1 > \lambda_2 > ... > \lambda_p > \lambda_{p+1} \geq ... \geq \lambda_n > 0$ are the eigenvalues of A, and $\boldsymbol{\phi}_i$ as the eigenvector corresponding to λ_i such that $\Phi = [\boldsymbol{\phi}_1 \,...\, \boldsymbol{\phi}_n]$ are orthonormal.

PF Deflation Algorithm

The objective function for the PF deflation PCA algorithm is

$$J\left(\mathbf{w}_k^i; A_k \right) = -\mathbf{w}_k^{i^T} A_k \mathbf{w}_k^i + \mu \left(\sum_{j=1}^{i-1} \left(\mathbf{w}_k^{j^T} \mathbf{w}_k^i \right)^2 + \frac{1}{2} \left(\mathbf{w}_k^{i^T} \mathbf{w}_k^i - 1 \right)^2 \right),\tag{5.18}$$

where $\mu > 0$ and $i=1,...,p$. The PF deflation adaptive gradient descent algorithm for PCA is

$$W_{k+1} = W_k + \eta_k \left(A_k W_k - \mu W_k \mathrm{UT} \left(W_k^T W_k - I_p \right) \right),\tag{5.19}$$

where UT[·] sets all elements below the diagonal of its matrix argument to zero. W_k converges to $W^* = [d_1\boldsymbol{\phi}_1 \, d_2\boldsymbol{\phi}_2 \,...\, d_p\boldsymbol{\phi}_p]$, where $d_i = \pm\sqrt{1 + (\lambda_i / \mu)}$.

PF Weighted Algorithm

The objective function for the PF weighted PCA algorithm is

$$J\left(\mathbf{w}_k^i; A_k\right) = -c_i \mathbf{w}_k^{i^T} A_k \mathbf{w}_k^i + \mu\left(\sum_{j=1, j\neq i}^{p} c_j \left(\mathbf{w}_k^{j^T} \mathbf{w}_k^i\right)^2 + \frac{c_i}{2}\left(\mathbf{w}_k^{i^T} \mathbf{w}_k^i - 1\right)^2 \right) \quad (5.20)$$

where $c_1 > c_2 > ... > c_p > 0$, $\mu > 0$, and $i=1,...,p$. The PF weighted adaptive gradient descent algorithm for PCA is

$$W_{k+1} = W_k + \eta_k \left(A_k W_k C - \mu W_k C\left(W_k^T W_k - I_p\right)\right), \quad (5.21)$$

where $C=diag(c_1,...,c_p)$. W_k converges to $W^* = [d_1\phi_1 \ d_2\phi_2 \ ... \ d_p\phi_p]$, where $d_i = \pm\sqrt{1+\left(\lambda_i / \mu\right)}$.

PF Algorithm Python Code

The following Python code implements the PF PCA algorithms with data X[nDim,nSamples]:

```
from numpy import linalg as la
nEA = 4 # number of PCA components computed
nEpochs  = 2
A  = np.zeros(shape=(nDim,nDim)) # stores adaptive
                              correlation matrix
W1 = 0.1 * np.ones(shape=(nDim,nEA)) # weight vectors of all
                              algorithms
W2 = W1
W3 = W1
c = [2-0.3*k for k in range(nEA)]
C = np.diag(c)
I  = np.identity(nEA)
mu = 2
```

```
for epoch in range(nEpochs):
    for iter in range(nSamples):
        cnt = nSamples*epoch + iter
        x = X[:,iter]
        x = x.reshape(nDim,1)
        A = A + (1.0/(1 + cnt))*((np.dot(x, x.T)) - A)
        # Homogeneous Gradient Descent
        W1 = W1 + (1/(100 + cnt))*(A @ W1 - mu * W1 @ ((W1.T
                                @ W1) - I))
        # Deflated Gradient Descent
        W2 = W2 + (1/(100 + cnt))*(A @ W2 - mu * W2
                                @ np.triu((W2.T @ W2) - I))
        # Weighted Gradient Descent
        W3 = W3 + (1/(100 + cnt))*(A @ W3 @ C - mu * W3 @ C
                                @ ((W3.T @ W3) - I))
```

5.6 AL1 Algorithms

AL1 Homogeneous Algorithm

We obtain the objective function for the augmented Lagrangian Method
1 homogeneous PCA algorithm by applying the augmented Lagrangian
method of nonlinear optimization to the Rayleigh quotient criterion as
follows:

$$J\left(\mathbf{w}_k^i; A_k\right) = -\mathbf{w}_k^{i^T} A_k \mathbf{w}_k^i + \alpha\left(\mathbf{w}_k^{i^T} \mathbf{w}_k^i - 1\right) + 2\sum_{j=1, j\neq i}^{p} \beta_j \mathbf{w}_k^{j^T} \mathbf{w}_k^i,$$

$$+\mu\left(\sum_{j=1, j\neq i}^{p}\left(\mathbf{w}_k^{j^T}\mathbf{w}_k^i\right)^2 + \frac{1}{2}\left(\mathbf{w}_k^{i^T}\mathbf{w}_k^i - 1\right)^2\right), \tag{5.22}$$

for $i=1,\ldots,p$, and $(\alpha,\beta_1,\beta_2,\ldots,\beta_p)$ are Lagrange multipliers and μ is a positive penalty constant. The objective function (5.22) is equivalent to $-tr\left(W_k^T A_k W_k\right)$ under the constraint $W_k^T W_k = I_p$, which also serves as the energy function for the AL1 algorithms.

Equating the gradient of (5.22) with respect to $\mathbf{w}_k^i$ to $\mathbf{0}$, and using the constraint $\mathbf{w}_k^{j^T}\mathbf{w}_k^i = \delta_{ij}$, we obtain

$$\alpha = \mathbf{w}_k^{i^T} A_k \mathbf{w}_k^i \text{ and } \beta_j = \mathbf{w}_k^{j^T} A_k \mathbf{w}_k^i \text{ for } j=1,\ldots,p,\, j\neq i. \tag{5.23}$$

Replacing $(\alpha,\beta_1,\beta_2,\ldots,\beta_p)$ in the gradient of (5.22), we obtain the AL1 homogeneous adaptive gradient descent algorithm for PCA:

$$W_{k+1} = W_k + \eta_k\left(\left(A_k W_k - W_k W_k^T A_k W_k - \mu W_k\left(W_k^T W_k - I_p\right)\right), \tag{5.24}$$

where $\mu>0$ and I_p is a $p\mathrm{X}p$ identity matrix. This algorithm is the same as the OJA algorithm (5.5) for $\mu=0$. We can prove that W_k converges $W^*=\Phi DU$, where $D=[D_1|0]^T\in\Re^{n\mathrm{X}p}$, $D_1= diag(d_1,\ldots,d_p)\in\Re^{p\mathrm{X}p}$, $d_i=\pm 1$ for $i=1,\ldots,p$, and $U\in\Re^{p\mathrm{X}p}$ is an arbitrary rotation matrix.

AL1 Deflation Algorithm

The objective function for the AL1 Deflation PCA algorithm is

$$J\left(\mathbf{w}_k^i;A_k\right) = -\mathbf{w}_k^{i^T} A_k \mathbf{w}_k^i + \alpha\left(\mathbf{w}_k^{i^T}\mathbf{w}_k^i - 1\right) + 2\sum_{j=1}^{i-1}\beta_j\mathbf{w}_k^{j^T}\mathbf{w}_k^i$$
$$+ \mu\left(\sum_{j=1}^{i-1}\left(\mathbf{w}_k^{j^T}\mathbf{w}_k^i\right)^2 + \frac{1}{2}\left(\mathbf{w}_k^{i^T}\mathbf{w}_k^i - 1\right)^2\right) \tag{5.25}$$

for $i=1,\ldots,p$. By solving for $(\alpha,\beta_1,\beta_2,\ldots,\beta_{i-1})$, and replacing them in the gradient of (5.25), we obtain

$$W_{k+1} = W_k + \eta_k\left(\left(A_k W_k - W_k\mathrm{UT}\left(W_k^T A_k W_k\right) - \mu W_k\mathrm{UT}\left(W_k^T W_k - I_p\right)\right), \tag{5.26}$$

where $\mu > 0$ and UT$[\cdot]$ sets all elements below the diagonal of its matrix argument to zero. W_k converges to $W^* = [\pm\phi_1 \pm\phi_2 \ldots \pm\phi_p]$ as $k \to \infty$.

AL1 Weighted Algorithm

The objective function for the AL1 weighted PCA algorithm is

$$J\left(\mathbf{w}_k^i; A_k\right) = -c_i {\mathbf{w}_k^i}^T A_k \mathbf{w}_k^i + \alpha c_i \left({\mathbf{w}_k^i}^T \mathbf{w}_k^i - 1\right) + 2 \sum_{j=1, j \neq i}^{p} \beta_j c_j {\mathbf{w}_k^j}^T \mathbf{w}_k^i,$$

$$+\mu \left(\sum_{j=1, j \neq i}^{p} c_j \left({\mathbf{w}_k^j}^T \mathbf{w}_k^i\right)^2 + \frac{c_i}{2} \left({\mathbf{w}_k^i}^T \mathbf{w}_k^i - 1\right)^2 \right) \qquad (5.27)$$

for $i = 1, \ldots, p$, where $(\alpha, \beta_1, \beta_2, \ldots, \beta_p)$ are Lagrange multipliers and μ is a positive penalty constant. By solving for $(\alpha, \beta_1, \beta_2, \ldots, \beta_p)$ and replacing them in the gradient of (5.27), we obtain the AL1 weighted adaptive gradient descent algorithm:

$$W_{k+1} = W_k + \eta_k \left(\left(A_k W_k C - W_k C W_k^T A_k W_k - \mu W_k C \left(W_k^T W_k - I_p \right) \right), \qquad (5.28)$$

where $\mu > 0$, $c_1 > c_2 > \ldots > c_p > 0$, and $C = diag(c_1, \ldots, c_p)$. We can prove that W_k converges to $W^* = [\pm\phi_1 \pm\phi_2 \ldots \pm\phi_p]$ as $k \to \infty$.

AL1 Algorithm Python Code

The following Python code implements the AL1 PCA algorithms with data X[nDim,nSamples]:

```
from numpy import linalg as la
nEA = 4 # number of PCA components computed
nEpochs  = 2
A  = np.zeros(shape=(nDim,nDim)) # stores adaptive
    correlation matrix
```

```python
W1 = 0.1 * np.ones(shape=(nDim,nEA)) # weight vectors of all
    algorithms
W2 = W1
W3 = W1
c = [2-0.3*k for k in range(nEA)]
C = np.diag(c)
I = np.identity(nEA)
mu = 2
for epoch in range(nEpochs):
    for iter in range(nSamples):
        cnt = nSamples*epoch + iter
        x = X[:,iter]
        x = x.reshape(nDim,1)
        A = A + (1.0/(1 + cnt))*((np.dot(x, x.T)) - A)
        # Homogeneous Gradient Descent
        W1 = W1 + (1/(100 + cnt))*(A @ W1 - W1 @ (W1.T @ A
                                  @ W1) -mu * W1 @ ((W1.T
                                  @ W1) - I))
        # Deflated Gradient Descent
        W2 = W2 + (1/(100 + cnt))*(A @ W2 - W2 @ np.triu(W2.T
                                  @ A @ W2) -mu * W2 @ np.triu
                                  ((W2.T @ W2) - I))
        # Weighted Gradient Descent
        W3 = W3 + (1/(300 + cnt))*(A @ W3 @ C - W3 @ C @ (W3.T
                                  @ A @ W3) -mu * W3 @ C @
                                  ((W3.T @ W3) - I))
```

5.7 AL2 Algorithms

AL2 Homogeneous Algorithm

The AL2 objective function can be derived from the AL1 homogeneous objective function (5.22) by replacing $\alpha, \beta_1, \beta_2, \ldots, \beta_p$ from (5.23) into (5.22) as

$$J\left(\mathbf{w}_k^i; A_k\right) = -\mathbf{w}_k^{i^T} A_k \mathbf{w}_k^i + \left(\mathbf{w}_k^{i^T} A_k \mathbf{w}_k^i\right)\left(\mathbf{w}_k^{i^T} \mathbf{w}_k^i - 1\right) + 2\sum_{j=1, j\neq i}^{p} \mathbf{w}_k^{i^T} A_k \mathbf{w}_k^j \mathbf{w}_k^{j^T} \mathbf{w}_k^i +$$
$$\mu\left(\sum_{j=1, j\neq i}^{p} \left(\mathbf{w}_k^{j^T} \mathbf{w}_k^i\right)^2 + \frac{1}{2}\left(\mathbf{w}_k^{i^T} \mathbf{w}_k^i - 1\right)^2\right), \tag{5.29}$$

for $i=1,\ldots,p$ and $\mu>0$. As seen with the XU objective function (5.10), (5.29) also has the constraints $\mathbf{w}_k^{i^T} \mathbf{w}_k^i = \delta_{ij}$ built into it. The AL2 homogeneous adaptive gradient descent algorithm for PCA is

$$W_{k+1} = W_k + \eta_k (2A_k W_k - W_k W_k^T A_k W_k - A_k W_k W_k^T W_k$$
$$- \mu W_k (W_k^T W_k - I_p)), \tag{5.30}$$

where I_p is a pXp identity matrix. W_k converges to $W^* = \Phi DU$, where $D=[D_1|0]^T \in \Re^{n\text{X}p}$, $D_1 = \mathrm{diag}(d_1,\ldots,d_p) \in \Re^{p\text{X}p}$, $d_i = \pm 1$ for $i=1,\ldots,p$, and $U \in \Re^{p\text{X}p}$ is an arbitrary rotation matrix.

AL2 Deflation Algorithm

The objective function for the AL2 deflation PCA algorithm is

$$J\left(\mathbf{w}_k^i; A_k\right) = -\mathbf{w}_k^{i^T} A_k \mathbf{w}_k^i + \left(\mathbf{w}_k^{i^T} A_k \mathbf{w}_k^i\right)\left(\mathbf{w}_k^{i^T} \mathbf{w}_k^i - 1\right) + 2\sum_{j=1}^{i-1} \mathbf{w}_k^{i^T} A_k \mathbf{w}_k^j \mathbf{w}_k^{j^T} \mathbf{w}_k^i +$$
$$\mu\left(\sum_{j=1}^{i-1}\left(\mathbf{w}_k^{j^T} \mathbf{w}_k^i\right)^2 + \frac{1}{2}\left(\mathbf{w}_k^{i^T} \mathbf{w}_k^i - 1\right)^2\right), \tag{5.31}$$

for $i=1,...,p$ and $\mu > 0$. Taking the gradient of (5.31) with respect to $\mathbf{w}_k^i$ we obtain the AL2 deflation adaptive gradient descent algorithm for PCA as

$$W_{k+1} = W_k + \eta_k (2A_k W_k - W_k \mathrm{UT}\left(W_k^T A_k W_k\right) - A_k W_k \mathrm{UT}\left(W_k^T W_k\right)$$
$$- \mu W_k \mathrm{UT}(W_k^T W_k - I_p)), \tag{5.32}$$

where $\mu > 0$, and UT$[\cdot]$ sets all elements below the diagonal of its matrix argument to zero. W_k converges to $W^* = [\pm\boldsymbol{\phi}_1 \pm\boldsymbol{\phi}_2 \ldots \pm\boldsymbol{\phi}_p]$ as $k\to\infty$.

AL2 Weighted Algorithm

The objective function for the AL2 weighted PCA algorithm is

$$J\left(\mathbf{w}_k^i; A_k\right) = -c_i \mathbf{w}_k^{i^T} A_k \mathbf{w}_k^i + c_i \left(\mathbf{w}_k^{i^T} A_k \mathbf{w}_k^i\right)\left(\mathbf{w}_k^{i^T} \mathbf{w}_k^i - 1\right)$$
$$+ 2 \sum_{j=1, j\neq i}^{p} c_j \mathbf{w}_k^{i^T} A_k \mathbf{w}_k^j \mathbf{w}_k^{j^T} \mathbf{w}_k^i +$$
$$\mu \left(\sum_{j=1, j\neq i}^{p} c_j \left(\mathbf{w}_k^{j^T} \mathbf{w}_k^i\right)^2 + \frac{c_i}{2}\left(\mathbf{w}_k^{i^T} \mathbf{w}_k^i - 1\right)^2\right), \tag{5.33}$$

where $i=1,...,p$, $\mu>0$, $c_1>c_2>...>c_p>0$. The AL2 weighted adaptive gradient descent algorithm is

$$W_{k+1} = W_k + \eta_k (2A_k W_k C - W_k C W_k^T A_k W_k - A_k W_k C W_k^T W_k$$
$$- \mu W_k C(W_k^T W_k - I_p)), \tag{5.34}$$

where $C=diag(c_1,...,c_p)$. W_k converges to $W^* = [\pm\boldsymbol{\phi}_1 \pm\boldsymbol{\phi}_2 \ldots \pm\boldsymbol{\phi}_p]$ as $k\to\infty$.

AL2 Algorithm Python Code

The following Python code implements the AL2 PCA algorithms with data X[nDim,nSamples]:

```python
from numpy import linalg as la
nEA = 4 # number of PCA components computed
nEpochs  = 2
A  = np.zeros(shape=(nDim,nDim)) # stores adaptive
                                 correlation matrix
W1 = 0.1 * np.ones(shape=(nDim,nEA)) # weight vectors of all
                                     algorithms
W2 = W1
W3 = W1
c = [2-0.3*k for k in range(nEA)]
C = np.diag(c)
I  = np.identity(nEA)
mu = 2
for epoch in range(nEpochs):
    for iter in range(nSamples):
        cnt = nSamples*epoch + iter
        x = X[:,iter]
        x = x.reshape(nDim,1)
        A = A + (1.0/(1 + cnt))*((np.dot(x, x.T)) - A)
        # Homogeneous Gradient Descent
        W1 = W1 + (1/(100 + cnt))*(A @ W1 - 0.5 * W1 @ (W1.T
                                   @ A @ W1) -0.5 * A @ W1
                                   @(W1.T @ W1) -0.5 * mu * W1
                                   @((W1.T @ W1) - I))
```

```
# Deflated Gradient Descent
W2 = W2 + (1/(100 + cnt))*(A @ W2 - 0.5 * W2 @ np.triu
                          (W2.T @ A @ W2) -0.5 * A
                          @ W2 @ np.triu(W2.T @ W2) -
                          0.5 * mu * W2 @ np.triu
                          ((W2.T @ W2) - I))
# Weighted Gradient Descent
W3 = W3 + (1/(100 + cnt))*(A @ W3 @ C - 0.5 * W3 @ C
                          @ (W3.T @ A @ W3) -
                          0.5 * A @ W3 @ C @ (W3.T
                          @ W3) -0.5 * mu * W3 @ C
                          @((W3.T @ W3) - I))
```

5.8 IT Algorithms

IT Homogeneous Function

The objective function for the information theory homogeneous PCA algorithm is

$$J\left(\mathbf{w}_k^i; A_k\right) = \mathbf{w}_k^{i^T} \mathbf{w}_k^i - \log\left(\mathbf{w}_k^{i^T} A_k \mathbf{w}_k^i\right) + \alpha\left(\mathbf{w}_k^{i^T} \mathbf{w}_k^i - 1\right)$$

$$+ 2 \sum_{j=1, j\neq i}^{p} \beta_j \mathbf{w}_k^{j^T} \mathbf{w}_k^i, \tag{5.35}$$

where $(\alpha, \beta_1, \beta_2, ..., \beta_p)$ are Lagrange multipliers and $i=1,...,p$. By equating the gradient of (5.35) with respect to $\mathbf{w}_k^i$ to $\mathbf{0}$, and using the constraint $\mathbf{w}_k^{j^T} \mathbf{w}_k^i = \delta_{ij}$, we obtain

$$\alpha = 0 \text{ and } \beta_j = \mathbf{w}_k^{j^T} A_k \mathbf{w}_k^i / \mathbf{w}_k^{i^T} A_k \mathbf{w}_k^i \text{ for } j=1,...,p, j\neq i. \tag{5.36}$$

Replacing $(\alpha, \beta_1, \beta_2, ..., \beta_p)$ in the gradient of (5.36), we obtain the IT homogeneous adaptive gradient descent algorithm for PCA:

$$W_{k+1} = W_k + \eta_k \left(A_k W_k - W_k W_k^T A_k W_k \right) \text{DIAG} \left(W_k^T A_k W_k \right)^{-1}, \qquad (5.37)$$

where DIAG[·] sets all elements *except* the diagonal of its matrix argument to zero, thereby making the matrix diagonal. W_k converges to $W^* = \Phi DU$, where $D = [D_1|0]^T \in \mathfrak{R}^{n \times p}$, $D_1 = diag(d_1, ..., d_p) \in \mathfrak{R}^{p \times p}$, $d_i = \pm 1$ for $i = 1, ..., p$, and $U \in \mathfrak{R}^{p \times p}$ is an arbitrary rotation matrix.

IT Deflation Algorithm

The objective function for the information theory deflation PCA algorithm is

$$J\left(\mathbf{w}_k^i; A_k \right) = \mathbf{w}_k^{i^T} \mathbf{w}_k^i - \log \left(\mathbf{w}_k^{i^T} A_k \mathbf{w}_k^i \right) + \alpha \left(\mathbf{w}_k^{i^T} \mathbf{w}_k^i - 1 \right) + 2 \sum_{j=1}^{i-1} \beta_j \mathbf{w}_k^{j^T} \mathbf{w}_k^i, \quad (5.38)$$

where $(\alpha, \beta_1, \beta_2, ..., \beta_{i-1})$ are Lagrange multipliers and $i = 1, ..., p$. By solving for $(\alpha, \beta_1, \beta_2, ..., \beta_{i-1})$ and replacing them in the gradient of (5.38), we obtain the IT deflation adaptive gradient descent algorithm for PCA:

$$W_{k+1} = W_k + \eta_k \left(A_k W_k - W_k \text{UT} \left(W_k^T A_k W_k \right) \right) \text{DIAG} \left(W_k^T A_k W_k \right)^{-1}. \qquad (5.39)$$

W_k converges with probability one to $W^* = [\pm\boldsymbol{\phi}_1 \pm\boldsymbol{\phi}_2 ... \pm\boldsymbol{\phi}_p]$ as $k \to \infty$.

IT Weighted Algorithm

The objective function for the IT weighted PCA algorithm is

$$J\left(\mathbf{w}_k^i; A_k \right) = c_i \mathbf{w}_k^{i^T} \mathbf{w}_k^i - c_i \log \left(\mathbf{w}_k^{i^T} A_k \mathbf{w}_k^i \right) + \alpha c_i \left(\mathbf{w}_k^{i^T} \mathbf{w}_k^i - 1 \right)$$

$$+ 2 \sum_{j=1, j\neq i}^{p} \beta_j c_j \mathbf{w}_k^{j^T} \mathbf{w}_k^i, \qquad (5.40)$$

for $i=1,\ldots,p$ and $(\alpha,\beta_1,\beta_2,\ldots,\beta_p)$ are Lagrange multipliers. The IT weighted adaptive gradient descent algorithm is

$$W_{k+1} = W_k + \eta_k \left(A_k W_k C - W_k C W_k^T A_k W_k \right) \mathrm{DIAG}\left(W_k^T A_k W_k \right)^{-1}, \qquad (5.41)$$

where $C=diag(c_1,\ldots,c_p)$ and $c_1>c_2>\ldots>c_p>0$. Here W_k converges to $W^*=[\pm\phi_1 \pm\phi_2 \ldots \pm\phi_p]$ as $k\to\infty$.

IT Algorithm Python Code

The following Python code implements the IT PCA algorithms with data X[nDim,nSamples]:

```python
from numpy import linalg as la
nEA = 4 # number of PCA components computed
nEpochs  = 3
A   = np.zeros(shape=(nDim,nDim)) # stores adaptive
                                correlation matrix
W1 = 0.1 * np.ones(shape=(nDim,nEA)) # weight vectors of all
                                algorithms
W2 = W1
W3 = W1
c = [2-0.3*k for k in range(nEA)]
C = np.diag(c)
for epoch in range(nEpochs):
    for iter in range(nSamples):
        cnt = nSamples*epoch + iter
        x = X[:,iter]
        x = x.reshape(nDim,1)
        A = A + (1.0/(1 + cnt))*((np.dot(x, x.T)) - A)
        # Homogeneous Gradient Descent
```

```
W1 = W1 + (1/(50 + cnt))*(A @ W1 - W1 @ (W1.T @ A
                        @ W1)) @ \inv(np.diag(
                        np.diagonal(W1.T @ A @ W1)))
# Deflated Gradient Descent
W2 = W2 + (1/(20 + cnt))*(A @ W2 - W2 @ np.triu(W2.T
                        @ A @ W2)) @ \inv(np.diag(
                        np.diagonal(W2.T @ A @ W2)))
# Weighted Gradient Descent
W3 = W3 + (1/(10 + cnt))*(A @ W3 @ C - W3 @ C @ (W3.T
                        @ A @ W3)) @ \inv(np.diag(
                        np.diagonal(W3.T @ A @ W3)))
```

5.9 RQ Algorithms

RQ Homogeneous Algorithm

We obtain the objective function for the Rayleigh quotient homogeneous PCA algorithm from the Rayleigh quotient as follows:

$$J\left(\mathbf{w}_k^i; A_k\right) = -\frac{\mathbf{w}_k^{i^T} A_k \mathbf{w}_k^i}{\mathbf{w}_k^{i^T} \mathbf{w}_k^i} + \alpha\left(\mathbf{w}_k^{i^T} \mathbf{w}_k^i - 1\right) + 2\sum_{j=1, j\neq i}^{p} \beta_j \mathbf{w}_k^{j^T} \mathbf{w}_k^i, \tag{5.42}$$

for $i=1,\ldots,p$ where $(\alpha,\beta_1,\beta_2,\ldots,\beta_p)$ are Lagrange multipliers. By equating the gradient of (5.42) with respect to $\mathbf{w}_k^i$ to $\mathbf{0}$, and using the constraint $\mathbf{w}_k^{j^T} \mathbf{w}_k^i = \delta_{ij}$, we obtain

$$\alpha = 0 \text{ and } \beta_j = \mathbf{w}_k^{j^T} A_k \mathbf{w}_k^i / \mathbf{w}_k^{i^T} \mathbf{w}_k^i \text{ for } j=1,\ldots,p, j\neq i. \tag{5.43}$$

Replacing $(\alpha,\beta_1,\beta_2,\ldots,\beta_p)$ in the gradient of (5.42) and making an approximation, we obtain the RQ homogeneous adaptive gradient descent algorithm for PCA:

$$W_{k+1} = W_k + \eta_k \left(A_k W_k - W_k W_k^T A_k W_k\right) \text{DIAG}\left(W_k^T W_k\right)^{-1}, \tag{5.44}$$

where DIAG[·] sets all elements except the diagonal of its matrix argument to zero. Here W_k converges to $W^* = \Phi DU$, where $D=[D_1|0]^T \in \mathfrak{R}^{n \times p}$, $D_1 = diag(d_v,...,d_p) \in \mathfrak{R}^{p \times p}$, $d_i = \pm 1$ for $i=1,...,p$, and $U \in \mathfrak{R}^{p \times p}$ is an arbitrary rotation matrix.

RQ Deflation Algorithm

The objective function for the RQ deflation PCA algorithm is

$$J\left(\mathbf{w}_k^i; A_k\right) = -\frac{\mathbf{w}_k^{i^T} A_k \mathbf{w}_k^i}{\mathbf{w}_k^{i^T} \mathbf{w}_k^i} + \alpha \left(\mathbf{w}_k^{i^T} \mathbf{w}_k^i - 1\right) + 2\sum_{j=1}^{i-1} \beta_j \mathbf{w}_k^{j^T} \mathbf{w}_k^i \qquad (5.45)$$

for $i=1,...,p$ where $(\alpha, \beta_1, \beta_2,..., \beta_{i-1})$ are Lagrange multipliers. By solving for $(\alpha, \beta_1, \beta_2,..., \beta_{i-1})$ and replacing them in the gradient of (5.45), we obtain the adaptive gradient descent algorithm:

$$W_{k+1} = W_k + \eta_k \left(A_k W_k - W_k \mathrm{UT}\left(W_k^T A_k W_k\right)\right) \mathrm{DIAG}\left(W_k^T W_k\right)^{-1}. \qquad (5.46)$$

W_k converges with probability one to $W^* = [\pm\boldsymbol{\phi}_1 \pm\boldsymbol{\phi}_2 ... \pm\boldsymbol{\phi}_p]$ as $k \to \infty$.

RQ Weighted Algorithm

The objective function for the RQ weighted PCA algorithm is

$$J\left(\mathbf{w}_k^i; A_k\right) = -c_i \frac{\mathbf{w}_k^{i^T} A_k \mathbf{w}_k^i}{\mathbf{w}_k^{i^T} \mathbf{w}_k^i} + \alpha c_i \left(\mathbf{w}_k^{i^T} \mathbf{w}_k^i - 1\right) + 2\sum_{j=1, j\neq i}^{p} \beta_j c_j \mathbf{w}_k^{j^T} \mathbf{w}_k^i, \qquad (5.47)$$

for $i=1,...,p$ and $(\alpha, \beta_1, \beta_2,..., \beta_p)$ are Lagrange multipliers. By solving for $(\alpha, \beta_1, \beta_2,..., \beta_p)$ and replacing them in the gradient of (5.47), we obtain the algorithm

$$W_{k+1} = W_k + \eta_k \left(A_k W_k C - W_k C W_k^T A_k W_k\right) \mathrm{DIAG}\left(W_k^T W_k\right)^{-1}, \qquad (5.48)$$

where $C = diag(c_1,...,c_p)$ and $c_1 > c_2 > ... > c_p > 0$. W_k converges with probability one to $W^* = [\pm\boldsymbol{\phi}_1 \pm\boldsymbol{\phi}_2 ... \pm\boldsymbol{\phi}_p]$ as $k \to \infty$.

RQ Algorithm Python Code

The following Python code implements the RQ PCA algorithms with data X[nDim,nSamples]:

```python
from numpy import linalg as la
nEA = 4 # number of PCA components computed
nEpochs  = 2
A  = np.zeros(shape=(nDim,nDim)) # stores adaptive
                               correlation matrix
W1 = 0.1 * np.ones(shape=(nDim,nEA))# weight vectors of all
                                   algorithms
W2 = W1
W3 = W1
c = [2-0.3*k for k in range(nEA)]
C = np.diag(c)
for epoch in range(nEpochs):
    for iter in range(nSamples):
        cnt = nSamples*epoch + iter
        x = X[:,iter]
        x = x.reshape(nDim,1)
        A = A + (1.0/(1 + cnt))*((np.dot(x, x.T)) - A)
        # Homogeneous Gradient Descent
        W1 = W1 + (1/(50 + cnt))*(A @ W1 - W1 @ (W1.T @ A @ W1)) @ \
                            inv(np.diag(np.diagonal(
                            W1.T @ W1)))
        # Deflated Gradient Descent
        W2 = W2 + (1/(20 + cnt))*(A @ W2 - W2 @ np.triu(W2.T
                                @ A @ W2)) @ \inv(np.diag(
                                np.diagonal(W2.T @ W2)))
```

```
# Weighted Gradient Descent
W3 = W3 + (1/(200 + cnt))*(A @ W3 @ C - W3 @ C @ (W3.T
                                @ A @ W3)) @ \inv(np.diag(
                                np.diagonal(W3.T @ W3)))
```

5.10 Summary of Adaptive Eigenvector Algorithms

I summarize the algorithms discussed here in Table 5-2. Each algorithm is of the form given in (5.1). The term $h(W_k, A_k)$ in (5.1) for each adaptive algorithm is given in Table 5-2. Note the following:

1. For each algorithm, I rate the compute and convergence performance.

2. I skip the homogeneous algorithms because they not useful for practical applications since they produce arbitrary rotations of the eigenvectors.

3. Note that $A_k \in \Re^{n \times n}$ and $W_k \in \Re^{n \times p}$. I present the computation complexity of each algorithm in terms of the matrix dimensions n and p.

4. The convergence performance is determined based on the speed of convergence of the principal and the minor components. I rate convergence on a scale of 1-10 where 10 is the best performing.

5. I skip the IT and RQ algorithms because they did not perform well compared to the remaining algorithms and the matrix inversion increases computational complexity.

Table 5-2. *List of Adaptive Eigen-Decomposition Algorithms*

Alg	Type	Adaptive Algorithm $h(W_k, A_k)$	Comment
OJA	Deflation	$A_k W_k - W_k \mathrm{UT}\left(W_k^T A_k W_k\right)$	$n^3 p^6$, 6
	Weighted	$A_k W_k C - W_k C W_k^T A_k W_k$	$n^4 p^6$, 6
XU	Deflation	$2 A_k W_k - A_k W_k \mathrm{UT}\left(W_k^T W_k\right)$ $- W_k \mathrm{UT}\left(W_k^T A_k W_k\right)$	$2 n^3 p^6$, 8
	Weighted	$2 A_k W_k C - W_k C W_k^T A_k W_k - A_k W_k C W_k^T W_k$	$2 n^4 p^6$, 8
PF	Deflation	$A_k W_k - \mu W_k \mathrm{UT}\left(W_k^T W_k - I_p\right)$	$n^2 p^4$, 7
	Weighted	$A_k W_k C - \mu W_k C\left(W_k^T W_k - I_p\right)$	$n^3 p^4$, 7
AL1	Deflation	$A_k W_k - W_k \mathrm{UT}\left(W_k^T A_k W_k\right)$ $- \mu W_k \mathrm{UT}\left(W_k^T W_k - I_p\right)$	$n^3 p^6 +$ $n^2 p^4$, 9
	Weighted	$A_k W_k C - W_k C W_k^T A_k W_k$ $- \mu W_k C\left(W_k^T W_k - I_p\right)$	$n^4 p^6 +$ $n^3 p^4$, 9
AL2	Deflation	$2 A_k W_k - W_k \mathrm{UT}\left(W_k^T A_k W_k\right)$ $- A_k W_k \mathrm{UT}\left(W_k^T W_k\right)$ $- \mu W_k \mathrm{UT}\left(W_k^T W_k - I_p\right)$	$2 n^3 p^6 +$ $n^2 p^4$, 10
	Weighted	$2 A_k W_k C - W_k C W_k^T A_k W_k$ $- A_k W_k C W_k^T W_k$ $- \mu W_k C\left(W_k^T W_k - I_p\right)$	$2 n^4 p^6 +$ $n^3 p^4$, 10

(continued)

Table 5-2. (*continued*)

Alg	Type	Adaptive Algorithm $h(W_k, A_k)$	Comment
IT	Deflation	$\left(A_k W_k - W_k \text{UT}\left(W_k^T A_k W_k\right)\right) \text{DIAG}\left(W_k^T A_k W_k\right)^{-1}$	Not applicable
	Weighted	$\left(A_k W_k C - W_k C W_k^T A_k W_k\right) \text{DIAG}\left(W_k^T A_k W_k\right)^{-1}$	Not applicable
RQ	Deflation	$\left(A_k W_k - W_k \text{UT}\left(W_k^T A_k W_k\right)\right) \text{DIAG}\left(W_k^T W_k\right)^{-1}$	Not applicable
	Weighted	$\left(A_k W_k C - W_k C W_k^T A_k W_k\right) \text{DIAG}\left(W_k^T W_k\right)^{-1}$	Not applicable

Observe the following:

1. The OJA algorithm has the least complexity and good performance.

2. The AL2 algorithm has the most complexity and best performance.

3. AL1 is the next best after AL2, and PF and Xu are the next best.

The complexity and accuracy tradeoffs will determine the algorithm to use in real-world scenarios. If you can afford the computation, the AL2 algorithm is the best. The XU algorithm is a good balance of complexity and speed of convergence.

5.11 Experimental Results

I generated 500 samples $\mathbf{x}_k$ of 10-dimensional Gaussian data (i.e., $n=10$) with the mean zero and covariance given below. The covariance matrix is obtained from the second covariance matrix in [Okada and Tomita 85] multiplied by 3. The covariance matrix is

$$3\begin{bmatrix}
0.427 & 0.011 & -0.005 & -0.025 & 0.089 & -0.079 & -0.019 & 0.074 & 0.089 & 0.005 \\
0.011 & 5.690 & -0.069 & -0.282 & -0.731 & 0.090 & -0.124 & 0.100 & 0.432 & -0.103 \\
-0.005 & -0.069 & 0.080 & 0.098 & 0.045 & -0.041 & 0.023 & 0.022 & -0.035 & 0.012 \\
-0.025 & -0.282 & 0.098 & 2.800 & -0.107 & 0.150 & -0.193 & 0.095 & -0.226 & 0.046 \\
0.089 & -0.731 & 0.045 & -0.107 & 3.440 & 0.253 & 0.251 & 0.316 & 0.039 & -0.010 \\
-0.079 & 0.090 & -0.041 & 0.150 & 0.253 & 2.270 & -0.180 & 0.295 & -0.039 & -0.113 \\
-0.019 & -0.124 & 0.023 & -0.193 & 0.251 & -0.180 & 0.327 & 0.027 & 0.026 & -0.016 \\
0.074 & 0.100 & 0.022 & 0.095 & 0.316 & 0.295 & 0.027 & 0.727 & -0.096 & -0.017 \\
0.089 & 0.432 & -0.035 & -0.226 & 0.039 & -0.039 & 0.026 & -0.096 & 0.715 & -0.009 \\
0.005 & -0.103 & 0.012 & 0.046 & -0.010 & -0.113 & -0.016 & -0.017 & -0.009 & 0.065
\end{bmatrix}.$$

The eigenvalues of the covariance matrix are

17.9013, 10.2212, 8.6078, 6.5361, 2.2396, 1.8369, 1.1361, 0.7693, 0.2245, 0.1503.

I computed the first four principal eigenvectors (i.e., the eigenvector corresponding to the largest four eigenvalues) (i.e., $p=4$) by the adaptive algorithms described here. In order to compute the online data sequence $\{A_k\}$, I generated random data vectors $\{\mathbf{x}_k\}$ from the above covariance matrix. I generated $\{A_k\}$ from $\{\mathbf{x}_k\}$ by using algorithm (2.5 in Section 2.4) with $\beta=1$. I computed the correlation matrix A after collecting all 500 samples $\mathbf{x}_k$ as

$$A = \frac{1}{500}\sum_{i=1}^{500}\mathbf{x}_i\mathbf{x}_i^T.$$

I referred to the eigenvectors and eigenvalues computed from this A by a standard numerical analysis method [Golub and VanLoan 83] as the *actual values*.

I started all algorithms with $\mathbf{w}_0 = 0.1*ONE$, where ONE is a 10X4 matrix whose all elements are ones. In order to measure the convergence and accuracy of the algorithms, I computed the direction cosine at k^{th} update of each adaptive algorithm as

$$\text{Direction cosine } (k) = \left| \mathbf{w}_k^{i\ T} \phi_i \right| \Big/ \| \phi_i \| \| \mathbf{w}_k^i \|, \tag{5.49}$$

where $\mathbf{w}_k^i$ is the estimated eigenvector of A_k at k^{th} update and ϕ_i is the actual i^{th} principal eigenvector computed from all collected samples by a conventional numerical analysis method.

Figure 5-3 shows the iterates of the OJA algorithms (deflated and weighted) to compute the first four principal eigenvectors of A. Figure 5-4 shows the same for the XU algorithms. Figure 5-5 shows the same for the PF algorithms. Figure 5-6 shows the iterates of the AL1 algorithms (deflated and weighted) to compute the first four principal eigenvectors of A. Figure 5-7 shows the same for the AL2 algorithms. Figure 5-8 shows the same for the IT algorithms. Figure 5-9 shows the same for the RQ algorithms.

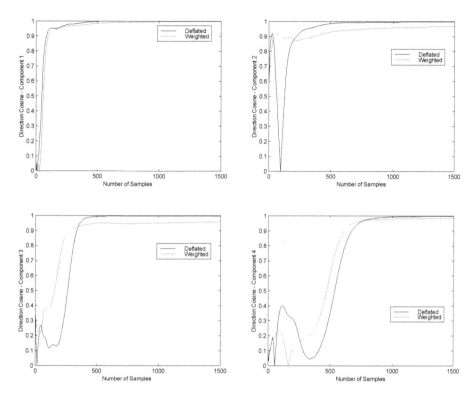

Figure 5-3. *Convergence of the first four principal eigenvectors of A by the OJA deflation (5.7) and OJA weighted (5.9) adaptive algorithms*

137

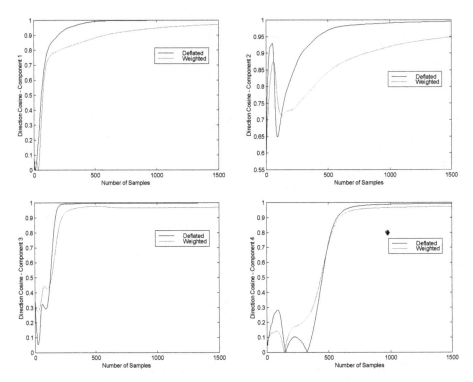

Figure 5-4. *Convergence of the first four principal eigenvectors of A by the XU deflation (5.13) and XU weighted (5.15) adaptive algorithms*

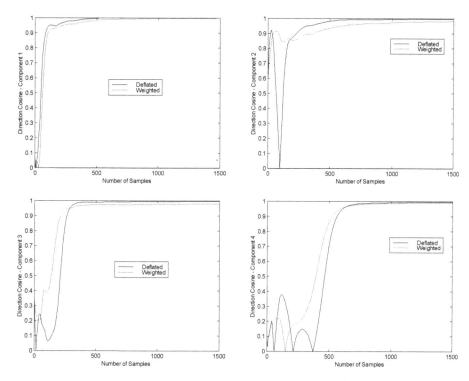

***Figure* 5-5.** *Convergence of the first four principal eigenvectors of A by the PF deflation (5.19) and PF weighted (5.21) adaptive algorithms*

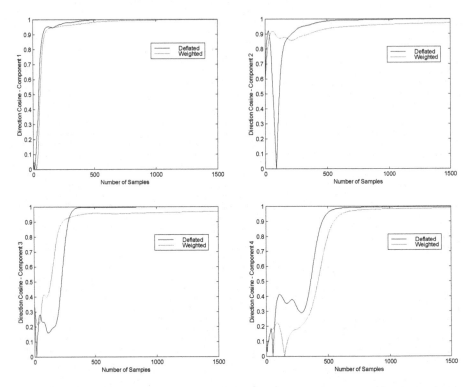

Figure 5-6. *Convergence of the first four principal eigenvectors of A by the AL1 deflation (5.26) and AL1 weighted (5.28) adaptive algorithms*

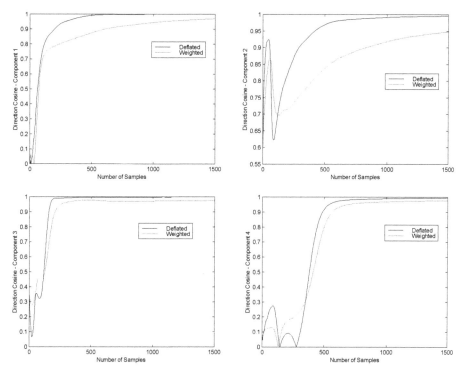

Figure 5-7. *Convergence of the first four principal eigenvectors of A by the AL2 deflation (5.32) and AL2 weighted (5.34) adaptive algorithms*

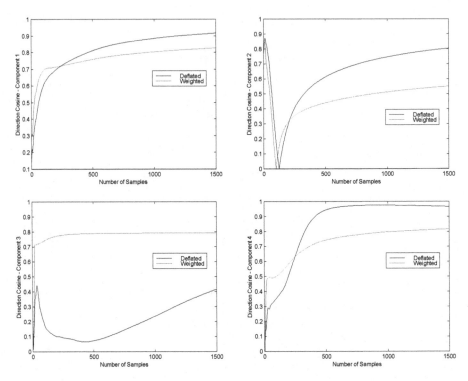

Figure 5-8. *Convergence of the first four principal eigenvectors of A by the IT deflation (5.39) and IT weighted (5.41) adaptive algorithms*

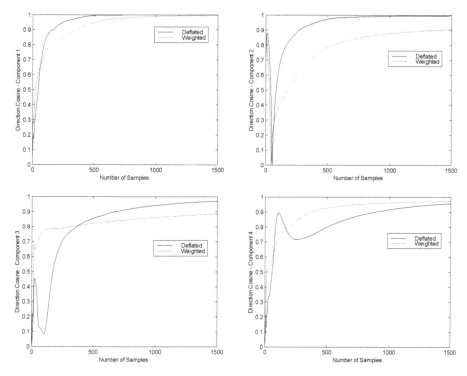

***Figure* 5-9.** *Convergence of the first four principal eigenvectors of A by the RQ deflation (5.46) and RQ weighted (5.48) adaptive algorithms*

The diagonal weight matrix C used for the weighted algorithms is DIAG(2.0, 1.7, 1.4, 1.1). I ran all algorithms for *three epochs* of the data, where one epoch means presenting all training data once in random order. I did not show the results for the homogeneous algorithms since the homogeneous method produces a linear combination of the actual eigenvectors of A. Thus, the direction cosines are not indicative of the performance of the algorithms for the homogeneous case.

5.12 Concluding Remarks

In this chapter, I discussed 21 different algorithms for adaptive PCA and viewed their convergence results. For each algorithm, I presented a common framework including an objective function from which I derived the adaptive algorithm. The deflation and weighted algorithms converged for all four principal eigenvectors, although the performance for the first principal eigenvector is the best. The convergences of all algorithms are similar, except for the IT algorithm, which did not perform as well as the rest.

Note that although I applied the gradient descent technique on the objective functions, I could have applied any other technique of nonlinear optimization such as steepest descent, conjugate direction, Newton-Raphson, or recursive least squares. The availability of the objective functions allows us to derive new algorithms by using new optimization techniques on them and also to perform convergence analyses of the adaptive algorithms.

CHAPTER 6

Accelerated Computation of Eigenvectors

6.1 Introduction

In Chapter 5, I discussed several adaptive algorithms for computing principal and minor eigenvectors of the online correlation matrix $A_k \in \mathfrak{R}^{n \times n}$ from a sequence of vectors $\{\mathbf{x}_k \in \mathfrak{R}^n\}$. I derived these algorithms by applying the gradient descent on an objective function. However, it is well known [Baldi and Hornik 95, Chatterjee et al. Mar 98, Haykin 94] that *principal component analysis* (PCA) algorithms based on gradient descents are slow to converge. Furthermore, both analytical and experimental studies show that convergence of these algorithms depends on appropriate selection of the gain sequence $\{\eta_k\}$. Moreover, it is proven [Chatterjee et al. Nov 97; Chatterjee et al. Mar 98; Chauvin 89] that if the gain sequence exceeds an upper bound, then the algorithms may diverge or converge to a false solution.

© Chanchal Chatterjee 2022
C. Chatterjee, *Adaptive Machine Learning Algorithms with Python*,
https://doi.org/10.1007/978-1-4842-8017-1_6

Since most of these algorithms are used for real-time (i.e., online) processing, it is especially difficult to determine an appropriate choice of the gain parameter at the start of the online process. Hence, it is important for wider applicability of these algorithms to

- Speed up the convergence of the algorithms, and

- Automatically select the gain parameter based on the current data sample.

Objective Functions for Gradient-Based Adaptive PCA

Some of the objective functions discussed in Chapters 4 and 5 have been used by practitioners to derive accelerated adaptive PCA algorithms by using advanced nonlinear optimization techniques such as

- Steepest descent (SD),

- Conjugate direction (CD),

- Newton-Raphson (NR), and

- Recursive least squares (RLS).

These optimization methods lead to faster convergence of the algorithms compared to the gradient descent methods discussed in Chapters 4 and 5. They also help us compute a value of the gain sequence $\{\eta_k\}$, which is not available in the gradient descent method. The only drawback is the additional computation needed by these improved algorithms. Note that each optimization method when applied to a different objective function leads to a new algorithm for adaptive PCA.

The first objective function that is used extensively in signal processing applications for accelerated PCA is the Rayleigh quotient (RQ) objective function described in Sections 4.4 and 5.9. Sarkar et al. [Sarkar et al. 89; Yang et al. 89] applied the steepest descent and conjugate direction methods to this objective function to compute the extremal (largest or smallest) eigenvectors. Fu and Dowling [Fu and Dowling 94, 95] generalized the conjugate direction algorithm to compute all eigenvectors. Zhu and Wang [Zhu and Wang 97] also used a conjugate direction method on a regularized total least squares version of this objective function. A survey of conjugate direction-based algorithms on the RQ objective function is found in [Yang et al. 89].

The second objective function that is used for accelerated PCA is the penalty function (PF) objective function discussed in Sections 4.7 and 5.5. Chauvin [Chauvin 97] presents a gradient descent algorithm based on the PF objective function and analyzes the landscape of this function. Mathew et al. [Mathew et al. 94, 95, 96] also use this objective function to offer a Newton-type algorithm for adaptive PCA.

The third objective function used for accelerated PCA is the information theoretic (IT) objective function discussed in Sections 4.5 and 5.8. Miao and Hua [Miao and Hua 98] present gradient descent and RLS algorithms for adaptive principal sub-space analysis.

The fourth objective function used for accelerated PCA is the XU objective function given in Sections 4.6 and 5.4. For example, Xu [Xu 93], Yang [Yang 95], Fu and Dowling [Fu and Dowling 94], Bannour and Azimi-Sadjadi [Bannour and Azimi-Sadjadi 95], and Miao and Hua [Miao and Hua 98] used variations of this objective function. As discussed in Section 5.4, there are several variations of this objective function including the mean squared error at the output of a two-layer linear auto-associative neural network. Xu derives an algorithm for adaptive principal sub-space analysis by using gradient descent. Yang uses gradient descent and recursive least squares optimization methods. Bannour and Azimi-Sadjadi also describe a recursive least squares-based algorithm for adaptive PCA with this objective function. Fu and Dowling reduce this objective function to one similar to the RQ objective function, which can be minimized by the conjugate direction methods due to Sarkar et al. [Sarkar et al. 89; Yang et al. 89]. They also compute the minor components by using an approximation and by employing the deflation technique.

Outline of This Chapter

Any of the objective functions discussed in Chapters 4 and 5 can be used to obtain accelerated adaptive PCA algorithms by using nonlinear optimization techniques (on this objective function) such as

- Gradient descent,

- Steepest descent,

- Conjugate direction,

- Newton-Raphson, and

- Recursive least squares.

I shall, however, use only one of these objective functions for the discussion in this chapter. I note that these analyses can be extended to the other objective functions in Chapters 4 and 5. My choice of objective function for this chapter is the XU deflation objective function discussed in Section 5.4.

Although gradient descent on the XU objective function (see Section 5.4) produces the well-known Xu's least mean square error reconstruction (LMSER) algorithm [Xu 93], the steepest descent, conjugate direction, and Newton-Raphson methods produce new adaptive algorithms for PCA [Chatterjee et al. Mar 00]. The penalty function (PF) deflation objective (see Section 5.5) function has also been accelerated by the steepest descent, conjugate direction, and quasi-Newton methods of optimization by Kang et al. [Kang et al. 00].

I shall apply these algorithms to stationary and **non-stationary multi-dimensional** Gaussian data sequences. I experimentally show that the adaptive steepest descent, conjugate direction, and Newton-Raphson algorithms converge much faster than the traditional gradient descent technique due to Xu [Xu 93]. Furthermore, the new algorithms automatically select the gain sequence $\{\eta_k\}$ based on the current data sample. I further compare the steepest descent algorithm with state-of-the-art methods such as Yang's Projection Approximation Subspace Tracking (PASTd) [Yang 95], Bannour and Sadjadi's recursive least squares (RLS) [Bannour et al. 95], and Fu and Dowling's conjugate gradient eigenstructure tracking (CGET1) [Fu and Dowling 94, 95] algorithms.

The XU deflation objective function for adaptive PCA algorithms is given in the Section 5.4 equation (5.12) as

$$J\left(\mathbf{w}_k^i; A_k\right) = -2\mathbf{w}_k^{i^T} A_k \mathbf{w}_k^i + \mathbf{w}_k^{i^T} A_k \mathbf{w}_k^i \mathbf{w}_k^{i^T} \mathbf{w}_k^i + 2\sum_{j=1}^{i-1} \mathbf{w}_k^{i^T} \mathbf{w}_k^j \mathbf{w}_k^{j^T} A_k \mathbf{w}_k^i, \quad (6.1)$$

for $i=1,...,p$, where $A_k \in \Re^{nXn}$ is the online observation matrix. I now apply different methods of nonlinear minimization to the objective function $J\left(\mathbf{w}_k^i; A_k\right)$ in (6.1) to obtain various algorithms for adaptive PCA.

In Sections 6.2, 6.3, 6.4, and 6.5, I apply the gradient descent, steepest descent, conjugate direction, and Newton-Raphson optimization methods to the unconstrained XU objective function for PCA given in (6.1). Here I obtain new algorithms for adaptive PCA. In Section 6.6, I present experimental results with stationary and non-stationary Gaussian sequences, thereby showing faster convergence of the new algorithms over traditional gradient descent adaptive PCA algorithms. I also compare the steepest descent algorithm with state-of-the-art algorithms. Section 6.7 concludes the chapter.

6.2 Gradient Descent Algorithm

[Chatterjee et al. *IEEE Trans. on Neural Networks*, Vol. 11, No. 2, pp. 338-355, March 2000.]

The gradient of (6.1) with respect to $\mathbf{w}_k^i$ is

$$\mathbf{g}_k^i = (1/2)\nabla_{\mathbf{w}_k^i} J\left(\mathbf{w}_k^i; A_k\right) = -2A_k\mathbf{w}_k^i + \sum_{j=1}^{i} A_k\mathbf{w}_k^j\mathbf{w}_k^{j^T}\mathbf{w}_k^i + \sum_{j=1}^{i} \mathbf{w}_k^j\mathbf{w}_k^{j^T} A_k\mathbf{w}_k^i$$

$$\text{for } i=1,\ldots,p. \tag{6.2}$$

Thus, the XU deflation adaptive gradient descent algorithm for PCA (see Section 5.4.2) is

$$\mathbf{w}_{k+1}^i = \mathbf{w}_k^i - \eta_k\mathbf{g}_k^i = \mathbf{w}_k^i + \eta_k\left(2A_k\mathbf{w}_k^i - \sum_{j=1}^{i} A_k\mathbf{w}_k^j\mathbf{w}_k^{j^T}\mathbf{w}_k^i - \sum_{j=1}^{i} \mathbf{w}_k^j\mathbf{w}_k^{j^T} A_k\mathbf{w}_k^i\right)$$

$$\text{for } i=1,\ldots,p, \tag{6.3}$$

where η_k is a decreasing gain constant. We can represent A_k simply by its instantaneous value $\mathbf{x}_k\mathbf{x}_k^T$ or by its recursive formula in Chapter 2 (Eq. 2.3 or 2.4). It is convenient to define a matrix $W_k = \left[\mathbf{w}_k^1 \ldots \mathbf{w}_k^p\right]$ ($p \leq n$), for which the columns are the p weight vectors that converge to the p principal eigenvectors of A_k respectively. Then, (6.2) can be represented as (same as (5.14) in Section 5.4.2):

$$W_{k+1} = W_k + \eta_k\left(2A_kW_k - A_kW_k\text{UT}\left(W_k^TW_k\right) - W_k\text{UT}\left(W_k^TA_kW_k\right)\right), \tag{6.4}$$

where UT[·] sets all elements below the diagonal of its matrix argument to zero, thereby making it upper triangular. Note that (6.2) is the LMSER algorithm due to Xu [Xu 93] that was derived from a least mean squared error criterion of a feed-forward neural network (see Section 5.4).

6.3 Steepest Descent Algorithm

[Chatterjee et al. *IEEE Trans. on Neural Networks*, Vol. 11, No. 2, pp. 338-355, March 2000.]

The adaptive steepest descent algorithm for PCA is obtained from $J\left(\mathbf{w}_k^i; A_k\right)$ in (6.1) as

$$\mathbf{w}_{k+1}^i = \mathbf{w}_k^i - \alpha_k^i \mathbf{g}_k^i, \tag{6.5}$$

where $\mathbf{g}_k^i$ is given in (6.2) and α_k^i is a non-negative scalar minimizing $J\left(\mathbf{w}_k^i - \alpha \mathbf{g}_k^i; A_k\right)$. Since we have an expression for $J\left(\mathbf{w}_k^i; A_k\right)$ in (6.1), we minimize the function $J\left(\mathbf{w}_k^i - \alpha \mathbf{g}_k^i; A_k\right)$ with respect to α and obtain the following cubic equation:

$$c_3 \alpha^3 + c_2 \alpha^2 + c_1 \alpha + c_0 = 0, \tag{6.6}$$

where

$$c_0 = -\mathbf{g}_k^{i^T} \mathbf{g}_k^i, \quad c_1 = \mathbf{g}_k^{i^T} H_k^i \mathbf{g}_k^i,$$

$$c_2 = -3\left(\mathbf{g}_k^{i^T} A_k \mathbf{g}_k^i \mathbf{w}_k^{i^T} \mathbf{g}_k^i + \mathbf{w}_k^{i^T} A_k \mathbf{g}_k^i \mathbf{g}_k^{i^T} \mathbf{g}_k^i\right), \quad c_3 = 2\mathbf{g}_k^{i^T} A_k \mathbf{g}_k^i \mathbf{g}_k^{i^T} \mathbf{g}_k^i.$$

Here H_k^i is the Hessian of $J\left(\mathbf{w}_k^i; A_k\right)$ as follows:

$$H_k^i = -2A_k + 2A_k \mathbf{w}_k^i \mathbf{w}_k^{i^T} + \mathbf{w}_k^i \mathbf{w}_k^{i^T} A_k + \mathbf{w}_k^{i^T} A_k \mathbf{w}_k^i I +$$
$$\sum_{j=1}^i A_k \mathbf{w}_k^j \mathbf{w}_k^{j^T} + \sum_{j=1}^i \mathbf{w}_k^j \mathbf{w}_k^{j^T} A_k. \tag{6.7}$$

With known values of $\mathbf{w}_k^i$ and $\mathbf{g}_k^i$, this cubic equation can be solved to obtain α that minimizes $J\left(\mathbf{w}_k^i - \alpha \mathbf{g}_k^i; A_k\right)$. A description of the computation of α is given in next section.

We now represent the adaptive PCA algorithm (6.5) in the matrix form. We define the matrices:

$$W_k = \left[\mathbf{w}_k^1, \ldots, \mathbf{w}_k^p\right], \ G_k = \left[\mathbf{g}_k^1, \ldots, \mathbf{g}_k^p\right], \text{ and } \Gamma_k = \text{Diag}\left(\alpha_k^1, \ldots, \alpha_k^p\right).$$

Then, the adaptive steepest descent PCA algorithm is

$$G_k = -2A_k W_k + W_k \text{UT}\left(W_k^T A_k W_k\right) + A_k W_k \text{UT}\left(W_k^T W_k\right),$$

$$W_{k+1} = W_k - G_k \Gamma_k. \tag{6.8}$$

Here UT[·] is the same as in (6.4).

Computation of α_k^i for Steepest Descent

From $J(\mathbf{w}_i; A)$ in (6.1), we compute α that minimizes $J(\mathbf{w}_i - \alpha \mathbf{g}_i; A)$, where

$$\mathbf{g}_i = -2A\mathbf{w}_i + \mathbf{w}_i\mathbf{w}_i^T A\mathbf{w}_i + \sum_{j=1}^{i-1}\mathbf{w}_j\mathbf{w}_j^T A\mathbf{w}_i + A\mathbf{w}_i\mathbf{w}_i^T\mathbf{w}_i + \sum_{j=1}^{i-1}A\mathbf{w}_j\mathbf{w}_j^T\mathbf{w}_i.$$

We have

$$\frac{dJ(\mathbf{w}_i - \alpha\mathbf{g}_i)}{d\alpha} = \frac{1}{2}tr\left[\nabla_{\mathbf{w}_i-\alpha\mathbf{g}_i} J(\mathbf{w}_i - \alpha\mathbf{g}_i)\frac{d(\mathbf{w}_i - \alpha\mathbf{g}_i)^T}{d\alpha}\right]$$

$$= -\frac{1}{2}\mathbf{g}_i^T\nabla_{\mathbf{w}_i-\alpha\mathbf{g}_i} J(\mathbf{w}_i - \alpha\mathbf{g}_i),$$

where

$$(1/2)\nabla_{\mathbf{w}_i-\alpha\mathbf{g}_i} J(\mathbf{w}_i - \alpha\mathbf{g}_i) = -2A(\mathbf{w}_i - \alpha\mathbf{g}_i)$$

$$+ (\mathbf{w}_i - \alpha\mathbf{g}_i)(\mathbf{w}_i - \alpha\mathbf{g}_i)^T A(\mathbf{w}_i - \alpha\mathbf{g}_i)$$

$$+ A(\mathbf{w}_i - \alpha\mathbf{g}_i)(\mathbf{w}_i - \alpha\mathbf{g}_i)^T (\mathbf{w}_i - \alpha\mathbf{g}_i) + \sum_{j=1}^{i-1}\mathbf{w}_j\mathbf{w}_j^T A(\mathbf{w}_i - \alpha\mathbf{g}_i)$$

$$+ \sum_{j=1}^{i-1}A\mathbf{w}_j\mathbf{w}_j^T (\mathbf{w}_i - \alpha\mathbf{g}_i).$$

Simplifying this equation, we obtain the following cubic equation:

$$c_3\alpha^3 + c_2\alpha^2 + c_1\alpha + c_0 = 0,$$

where

$$c_0 = -\mathbf{g}_i^T\mathbf{g}_i, \ c_1 = \mathbf{g}_i^T H_i\mathbf{g}_i,$$

$$c_2 = -3\left(\mathbf{g}_i^T A\mathbf{g}_i\mathbf{w}_i^T\mathbf{g}_i + \mathbf{w}_i^T A\mathbf{g}_i\mathbf{g}_i^T\mathbf{g}_i\right), \ c_3 = 2\mathbf{g}_i^T A\mathbf{g}_i\mathbf{g}_i^T\mathbf{g}_i.$$

Here H_i is the Hessian of $J(\mathbf{w}_i; A)$ given in (6.7).

It is well known that a cubic polynomial has at least one real root (two complex conjugate roots with a real root or three real roots). The roots can also be computed in closed form as shown in [Artin 91]. If a root is complex, then $\mathbf{w}_i - \alpha \mathbf{g}_i$ is complex, and clearly this is not the root we are looking for. If we have three real roots, then we can either take the root corresponding to minimum $J(\mathbf{w}_i - \alpha \mathbf{g}_i; A)$ or the one corresponding to $3c_3\alpha^2 + 2c_2\alpha + c_1 > 0$.

Steepest Descent Algorithm Code

The following Python code implements this algorithm with data X[nDim,nSamples]:

```python
from numpy import linalg as la
A  = np.zeros(shape=(nDim,nDim)) # stores adaptive
                                 correlation matrix
W1 = 0.1 * np.ones(shape=(nDim,nEA)) # weight vectors of all
                                     algorithms
W2 = W1
I  = np.identity(nDim)
Weight = 1
nEpochs = 1
for epoch in range(nEpochs):
    for iter in range(nSamples):
        cnt = nSamples*epoch + iter
        # Update data correlation matrix A with current
        sample x
        x = X1[:,iter]
        x = x.reshape(nDim,1)
        A = Weight * A + (1.0/(1 + cnt))*((np.dot(x, x.T)) -
            Weight * A)
```

```python
# Steepest Descent
G = -2 * A @ W2 + A @ W2 @ np.triu(W2.T @ W2) + \
    W2 @ np.triu(W2.T @ A @ W2)
for i in range(nEA):
    M = np.zeros(shape=(nDim,nDim))
    for k in range(i):
        wk = W2[:,k].reshape(nDim,1)
        M = M + (A @ (wk @ wk.T) + (wk @ wk.T) @ A)
    wi = W2[:,i].reshape(nDim,1)
    F = - 2*A + 2*A @ (wi @ wi.T) + 2 * (wi
        @ wi.T) @ A + \
        A * (wi.T @ wi) + (wi.T @ A @ wi) * I  +  M
    gi = G[:,i].reshape(nDim,1)
    a0 = np.asscalar(gi.T @ gi)
    a1 = np.asscalar(- gi.T @ F @ gi)
    a2 = np.asscalar(3 * ((wi.T @ A @ gi) @ (gi.T
                             @ gi) + \
                          (gi.T @ A @ gi)*(wi.T @ gi)))
    a3 = np.asscalar(- 2 * (gi.T @ A @ gi)
                     @ (gi.T @ gi))
    c  = np.array([a3, a2, a1, a0])
    rts = np.roots(c)
    rs = np.zeros(3)
    r  = np.zeros(3)
    J  = np.zeros(3)
    cnt1 = 0
    for k in range(3):
        if np.isreal(rts[k]):
            re = np.real(rts[k])
            rs[cnt1] = re
            r = W2[:,i] - re * G[:,i]
```

```
J[cnt1] = np.asscalar(-2*(r.T @ A @ r) +
                        (r.T @ A @ r) * \
                        (r.T @ r) + (r.T
                        @ M @ r))
        cnt1 = cnt1 + 1
yy = min(J)
iyy = np.argmin(J)
alpha = rs[iyy]
W2[:,i] = (W2[:,i] - alpha * G[:,i]).T
```

6.4 Conjugate Direction Algorithm

[Chatterjee et al. *IEEE Trans. on Neural Networks*, Vol. 11, No. 2, pp. 338-355, March 2000.]

The adaptive conjugate direction algorithm for PCA can be obtained as follows:

$$\mathbf{w}^i_{k+1} = \mathbf{w}^i_k + \alpha^i_k \mathbf{d}^i_k$$

$$\mathbf{d}^i_{k+1} = -\mathbf{g}^i_{k+1} + \beta^i_k \mathbf{d}^i_k,\tag{6.9}$$

where $\mathbf{g}^i_{k+1} = (1/2)\nabla_{\mathbf{w}_i} J\left(\mathbf{w}^i_{k+1}; A_k\right)$. The gain constant α^i_k is chosen as α that minimizes $J\left(\mathbf{w}^i_k + \alpha \mathbf{d}^i_k\right)$. Similar to the steepest descent case, we obtain the following cubic equation:

$$c_3\alpha^3 + c_2\alpha^2 + c_1\alpha + c_0 = 0,\tag{6.10}$$

where

$$c_0 = \mathbf{g}^i_k{}^T \mathbf{d}^i_k,\ c_1 = \mathbf{d}^i_k{}^T H^i_k \mathbf{d}^i_k,$$

$$c_2 = 3\left(\mathbf{d}^i_k{}^T A_k \mathbf{d}^i_k \mathbf{w}^i_k{}^T \mathbf{d}^i_k + \mathbf{w}^i_k{}^T A_k \mathbf{d}^i_k \mathbf{d}^i_k{}^T \mathbf{d}^i_k\right),\ c_3 = 2\mathbf{d}^i_k{}^T A_k \mathbf{d}^i_k \mathbf{d}^i_k{}^T \mathbf{d}^i_k.$$

Here, $\mathbf{g}_k^i = (1/2)\nabla_{\mathbf{w}_i} J\left(\mathbf{w}_k^i; A_k\right)$ as given in (6.2). Equation (6.10) is solved to obtain α that minimizes $J\left(\mathbf{w}_k^i + \alpha \mathbf{d}_k^i\right)$. For the choice of β_k^i, we can use a number of methods such as Hestenes-Stiefel, Polak-Ribiere, Fletcher-Reeves, and Powell (described on Wikipedia).

We now represent the adaptive conjugate direction PCA algorithm (6.9) in the matrix form. We define the following matrices:

$$W_k = \left[\mathbf{w}_k^1, \ldots, \mathbf{w}_k^p\right], \; G_k = \left[\mathbf{g}_k^1, \ldots, \mathbf{g}_k^p\right], \; D_k = \left[\mathbf{d}_k^1, \ldots, \mathbf{d}_k^p\right],$$

$$\Gamma_k = \operatorname{diag}\left(\alpha_k^1, \ldots, \alpha_k^p\right), \text{ and } \Pi_k = \operatorname{diag}\left(\beta_k^1, \ldots, \beta_k^p\right).$$

Then, the adaptive conjugate direction PCA algorithm is

$$W_{k+1} = W_k + D_k \Gamma_k,$$

$$G_{k+1} = -2A_k W_{k+1} + W_{k+1} \mathrm{UT}\left(W_{k+1}^T A_k W_{k+1}\right) + A_k W_{k+1} \mathrm{UT}\left(W_{k+1}^T W_{k+1}\right),$$

$$D_{k+1} = -G_{k+1} + D_k \Pi_k. \tag{6.11}$$

Here UT[·] is the same as in (6.4).

Conjugate Direction Algorithm Code

The following Python code implements this algorithm with data X[nDim,nSamples]:

```
from numpy import linalg as la
A  = np.zeros(shape=(nDim,nDim)) # stores adaptive
                                 correlation matrix
W1 = 0.1 * np.ones(shape=(nDim,nEA)) # weight vectors of all
                                     algorithms

W2 = W1
I  = np.identity(nDim)
Weight = 1
```

```python
nEpochs = 1
for epoch in range(nEpochs):
    for iter in range(nSamples):
        cnt = nSamples*epoch + iter

        # Update data correlation matrix A with current
        sample x
        x = X1[:,iter]
        x = x.reshape(nDim,1)
        A = Weight * A + (1.0/(1 + cnt))*((np.dot(x, x.T)) -
            Weight * A)

        # Conjugate Direction Method
        # Initialize D
        G = -2 * A @ W2 + A @ W2 @ np.triu(W2.T @ W2) +
                \W2 @ np.triu(W2.T @ A @ W2)
        if (iter == 0):
            D = G
        # Update W
        for i in range(nEA):
            gi = G[:,i].reshape(nDim,1)
            wi = W2[:,i].reshape(nDim,1)
            di = D[:,i].reshape(nDim,1)
            M = np.zeros(shape=(nDim,nDim))
            for k in range(i):
                wk = W2[:,k].reshape(nDim,1)
                M = M + (A @ (wk @ wk.T) + (wk @ wk.T) @ A)
            F = - 2*A + 2*A @ (wi @ wi.T) + 2 * (wi @ wi.T) @
                A + A * \
                    (wi.T @ wi) + (wi.T @ A @ wi) * I  +  M
            a0 = np.asscalar(gi.T @ di)
            a1 = np.asscalar(- di.T @ F @ di)
```

```python
        a2 = np.asscalar(3 * ((wi.T @ A @ di) * (di.T
                        @ di) + \
                              (di.T @ A @ di) *
                              (wi.T @ di)))
        a3 = np.asscalar(- 2 * (di.T @ A @ di) *
                        (di.T @ di))
        c  = np.array([a3, a2, a1, a0])
        rts = np.roots(c)
        rs = np.zeros(3)
        r  = np.zeros(3)
        J  = np.zeros(3)
        cnt1 = 0
        for k in range(3):
            if np.isreal(rts[k]):
                re = np.real(rts[k])
                rs[cnt1] = re
                r = (W2[:,i] - re * di.T).reshape(nDim,1)
                J[cnt1] = np.asscalar(-2*(r.T @ A @ r) +
                        (r.T @ A @ r) * \
                                    (r.T @ r) + (r.T
                                    @ M @ r))
                cnt1 = cnt1 + 1
        yy = min(J)
        iyy = np.argmin(J)
        alpha = rs[iyy]
        W2[:,i] = W2[:,i] - alpha * di.T
        # Update d
        gi = G[:,i].reshape(nDim,1)
        wi = W2[:,i].reshape(nDim,1)
        di = D[:,i].reshape(nDim,1)
        M = np.zeros(shape=(nDim,nDim))
```

```
for k in range(i):
    wk = W2[:,k].reshape(nDim,1)
    M = M + (A @ (wk @ wk.T) + (wk @ wk.T) @ A)
F = - 2*A + 2*A @ (wi @ wi.T) + 2 * (wi @
    wi.T) @ A + \
        A * (wi.T @ wi) + (wi.T @ A @ wi) * I  +  M
beta = (gi.T @ F @ di) / (di.T @ F @ di)
di = gi + 1*beta*di
D[:,i] = di.T
```

6.5 Newton-Raphson Algorithm

[Chatterjee et al. *IEEE Trans. on Neural Networks*, Vol. 11, No. 2, pp. 338-355, March 2000.]

The adaptive Newton-Raphson algorithm for PCA is

$$\mathbf{w}_{k+1}^i = \mathbf{w}_k^i - \alpha_k^i \left(H_k^i \right)^{-1} \mathbf{g}_k^i,$$

(6.12)

where α_k^i is a non-negative scalar and H_k^i is the online Hessian given in (6.7). The search parameter α_k^i is commonly selected: (1) by minimizing $\cdot J\left(\mathbf{w}_k^i + \alpha \mathbf{d}_k^i \right)$ where $\mathbf{d}_k^i = -\left(H_k^i \right)^{-1} \mathbf{g}_k^i$; (2) as a scalar constant; or (3) as a decreasing sequence $\{\alpha_k^i\}$ such that $\alpha_k^i \to 0$ as $k \to \infty$.

The main concerns in this algorithm are that H_k^i should be positive definite, and that we should adaptively obtain an estimate of $\left(H_k^i \right)^{-1}$ in order to make the algorithm computationally efficient. These two concerns are addressed if we approximate the Hessian by dropping the term $\left(-A_k + A_k \mathbf{w}_k^{i^T} \mathbf{w}_k^i \right)$, which is close to 0 for $\mathbf{w}_k^i$ close to the solution. The new Hessian is

$$H_k^i \approx \mathbf{w}_k^{i^T} A_k \mathbf{w}_k^i I - \tilde{A}_k^i + 2 A_k \mathbf{w}_k^i \mathbf{w}_k^{i^T} + 2 \mathbf{w}_k^i \mathbf{w}_k^{i^T} A_k,$$

(6.13)

where

$$\tilde{A}_k^i = A_k - \sum_{j=1}^{i-1} \mathbf{w}_k^j \mathbf{w}_k^{j^T} A_k - \sum_{j=1}^{i-1} A_k \mathbf{w}_k^j \mathbf{w}_k^{j^T}.$$

We can compute A_k^i by an iterative equation in i as follows:

$$\tilde{A}_k^{i+1} = \tilde{A}_k^i - \mathbf{w}_k^i \mathbf{w}_k^{i^T} A_k - A_k \mathbf{w}_k^i \mathbf{w}_k^{i^T}.$$

Inverting this Hessian consists of inverting the matrix $B_k^i = \mathbf{w}_k^{i^T} A_k \mathbf{w}_k^i I - \tilde{A}_k^i$ and two rank-one updates. An approximate inverse of this matrix B_k^i is given by

$$\left(B_k^i\right)^{-1} = \left(\mathbf{w}_k^{i^T} A_k \mathbf{w}_k^i I - \tilde{A}_k^i\right)^{-1} \approx \frac{I + \tilde{A}_k^i / \mathbf{w}_k^{i^T} A_k \mathbf{w}_k^i}{\mathbf{w}_k^{i^T} A_k \mathbf{w}_k^i}. \tag{6.14}$$

An adaptive algorithm for inverting the Hessian H_k^i in (6.13) can be obtained by two rank-one updates. Let's define

$$C_k^i = B_k^i + 2 A_k \mathbf{w}_k^i \mathbf{w}_k^{i^T}. \tag{6.15}$$

Then from (6.13), an update formula for $\left(H_k^i\right)^{-1}$ is

$$\left(H_k^i\right)^{-1} = \left(C_k^i\right)^{-1} - \frac{2\left(C_k^i\right)^{-1} \mathbf{w}_k^i \mathbf{w}_k^{i^T} A_k \left(C_k^i\right)^{-1}}{1 + 2\mathbf{w}_k^{i^T} A_k \left(C_k^i\right)^{-1} \mathbf{w}_k^i}, \tag{6.16}$$

where $\left(C_k^i\right)^{-1}$ is obtained from (6.15) as

$$\left(C_k^i\right)^{-1} = \left(B_k^i\right)^{-1} - \frac{2\left(B_k^i\right)^{-1} A_k \mathbf{w}_k^i \mathbf{w}_k^{i^T} \left(B_k^i\right)^{-1}}{1 + 2\mathbf{w}_k^{i^T} \left(B_k^i\right)^{-1} A_k \mathbf{w}_k^i} \tag{6.17}$$

and $\left(B_k^i\right)^{-1}$ is given in (6.14).

Newton-Raphson Algorithm Code

The following Python code implements this algorithm with data
X[nDim,nSamples]:

```python
from numpy import linalg as la
A  = np.zeros(shape=(nDim,nDim)) # stores adaptive
                                 correlation matrix
W1 = 0.1 * np.ones(shape=(nDim,nEA)) # weight vectors of all
                                     algorithms
W2 = W1
I  = np.identity(nDim)
Weight = 1
nEpochs = 1
for epoch in range(nEpochs):
    for iter in range(nSamples):
        cnt = nSamples*epoch + iter
        # Update data correlation matrix A with current sample x
        x = X1[:,iter]
        x = x.reshape(nDim,1)
        A = Weight * A + (1.0/(1 + cnt))*((np.dot(x, x.T)) -
            Weight * A)
        # Newton Rhapson
        G = -2* A @ W4 + A@ W4@ np.triu(W4.T @ W4) + W4@
            np.triu(W4.T @ A @ W4)
        # Update W
        for i in range(nEA):
            M = np.zeros(shape=(nDim,nDim))
            for k in range(i):
                wk = W4[:,k].reshape(nDim,1)
                M = M + (A @ (wk @ wk.T) + (wk @ wk.T) @ A)
            wi = W4[:,i].reshape(nDim,1)
```

```python
F = - 2*A + 2*A @ (wi @ wi.T) + 2 * (wi @ wi.T) @ A + \
    A * (wi.T @ wi) + (wi.T @ A @ wi) * I  +  M
lam = wi.T @ A @ wi
Atilde = A
if (iter > 0):
    invB = (I + (Atilde/lam))/lam
invC = invB- (2*invB@ A @wi @wi.T @ invB)/
(1 + 2*wi.T@ invB@ A @ wi)
invF = invC- (2*invC@ wi @wi.T @A @ invC)/
(1 + 2*wi.T@ A @invC @ wi)
gi = G[:,i].reshape(nDim,1)
di = -invF @ gi
a0 = np.asscalar(gi.T @ di)
a1 = np.asscalar(di.T @ F @ di)
a2 = np.asscalar(3* ((wi.T @ A @ di) @ (di.T @ di) + \
                (di.T @ A @ di)*(wi.T @ di)))
a3 = np.asscalar(2 * (di.T @ A @ di) @ (di.T @ di))
c  = np.array([a3, a2, a1, a0])
rts = np.roots(c)
rs = np.zeros(3)
r  = np.zeros(3)
J  = np.zeros(3)
cnt1 = 0
for k in range(3):
    if np.isreal(rts[k]):
        re = np.real(rts[k])
        rs[cnt1] = re
        r = W4[:,i] + re * di.reshape(nDim)
        J[cnt1] = np.asscalar(-2*(r.T @ A @ r) +
                (r.T @ A @ r) * \
                            (r.T @ r) + (r.T
                            @ M @ r))
```

```
                  cnt1 = cnt1 + 1
      iyy = np.argmin(J)
      alpha = rs[iyy]
      W4[:,i] = (W4[:,i] + alpha * di.reshape(nDim)).T
```

6.6 Experimental Results

I did two sets of experiments to test the performance of the accelerated PCA algorithms. I did the first set of experiments on *stationary* Gaussian data and the second set on *non-stationary* Gaussian data. I then compared the steepest descent algorithm against state-of-the-art adaptive PCA algorithms like Yang's Projection Approximation Subspace Tracking, Bannour and Sadjadi's Recursive Least Squares, and Fu and Dowling's Conjugate Gradient Eigenstructure Tracking algorithms.

Experiments with Stationary Data

I generated 2000 samples of 10-dimensional Gaussian data (i.e., n=10) with mean zero and covariance given below. Note that this covariance matrix is obtained from the first covariance matrix in [Okada and Tomita 85] multiplied by 2. The covariance matrix is

$$2\begin{bmatrix} 0.091 & 0.038 & -0.053 & -0.005 & 0.010 & -0.136 & 0.155 & 0.030 & 0.002 & 0.032 \\ 0.038 & 0.373 & 0.018 & -0.028 & -0.011 & -0.367 & 0.154 & -0.057 & -0.031 & -0.065 \\ -0.053 & 0.018 & 1.430 & 0.017 & 0.055 & -0.450 & -0.038 & -0.298 & -0.041 & -0.030 \\ -0.005 & -0.028 & 0.017 & 0.084 & -0.005 & 0.016 & 0.042 & -0.022 & 0.001 & 0.005 \\ 0.010 & -0.011 & 0.055 & -0.005 & 0.071 & 0.088 & 0.058 & -0.069 & -0.008 & 0.003 \\ -0.136 & -0.367 & -0.450 & 0.016 & 0.088 & 5.720 & -0.544 & -0.248 & 0.005 & 0.095 \\ 0.155 & 0.154 & -0.038 & 0.042 & 0.058 & -0.544 & 2.750 & -0.343 & -0.011 & -0.120 \\ 0.030 & -0.057 & -0.298 & -0.022 & -0.069 & -0.248 & -0.343 & 1.450 & 0.078 & 0.028 \\ 0.002 & -0.031 & -0.041 & 0.001 & -0.008 & 0.005 & -0.011 & 0.078 & 0.067 & 0.015 \\ 0.032 & -0.065 & -0.030 & 0.005 & 0.003 & 0.095 & -0.120 & 0.028 & 0.015 & 0.341 \end{bmatrix}.$$

The eigenvalues of the covariance matrix are

11.7996, 5.5644, 3.4175, 2.0589, 0.7873, 0.5878, 0.1743, 0.1423, 0.1213, 0.1007.

Clearly, the first four eigenvalues are significant and I adaptively compute the corresponding eigenvectors (i.e., p=4). See Figure 6-1 for the plots of the 10-dimensional random stationary data.

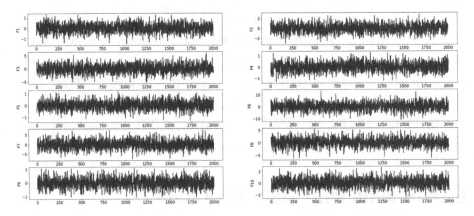

Figure 6-1. *10-dimensional stationary random normal data*

In order to compute the online data sequence $\{A_k\}$, I generated random data vectors $\{\mathbf{x}_k\}$ from the above covariance matrix. I generated $\{A_k\}$ from $\{\mathbf{x}_k\}$ by using algorithm (2.3) in Chapter 2. I computed the correlation matrix A after collecting all 500 samples $\mathbf{x}_k$ as

$$A = \frac{1}{2000}\sum_{i=1}^{2000}\mathbf{x}_i\mathbf{x}_i^T.$$

I refer to the eigenvectors and eigenvalues computed from this A by a standard numerical analysis method [Golub and VanLoan 83] as the *actual values*.

I used the adaptive gradient descent (6.4), steepest descent (6.8), conjugate direction (6.11), and Newton-Raphson (6.12) algorithms on the random data sequence $\{A_k\}$. I started all algorithms with $W_0 = 0.1*ONE$, where ONE is a 10 X 4 matrix whose all elements are ones. In order to measure the convergence and accuracy of the algorithms, I computed the direction cosine at k^{th} update of each adaptive algorithm as

$$\text{Direction cosine } (k) = \frac{\| \mathbf{w}_k^{i\,T} \boldsymbol{\phi}_i \|}{\| \mathbf{w}_k^i \| \| \boldsymbol{\phi}_i \|}, \tag{6.18}$$

where $\mathbf{w}_k^i$ is the estimated eigenvector of A_k at k^{th} update and $\boldsymbol{\phi}_i$ is the actual eigenvector computed from all collected samples by a conventional numerical analysis method.

Figures 6-2 through 6-4 show the iterates of the four algorithms to compute the first four principal eigenvectors of A. For the gradient descent (6.4) algorithm, I used $\eta_k=1/(400+k)$. For the conjugate direction method, I used the Hestenes-Stiefel [Nonlinear conjugate gradient method, Wikipedia] method (see Section 6.4) to compute β_k^i. For the steepest descent, conjugate direction, and Newton-Raphson methods, I chose α_k^i by solving a cubic equation as described in Sections 6.3, 6.4, and 6.5, respectively.

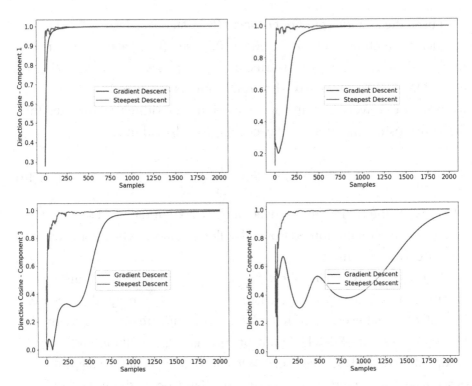

Figure 6-2. *Convergence of the first four principal eigenvectors of A by the gradient descent (6.4) and steepest descent (6.8) algorithms for stationary data*

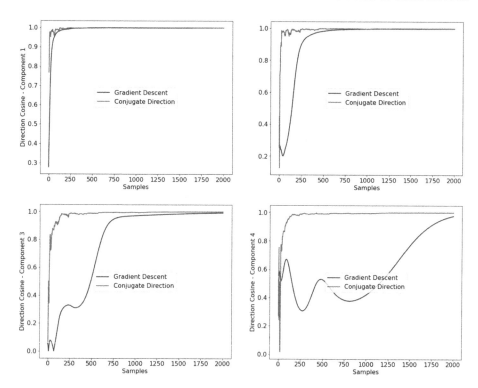

Figure 6-3. *Convergence of the first four principal eigenvectors of A by the gradient descent (6.4) and conjugate direction (6.11) algorithms for stationary data*

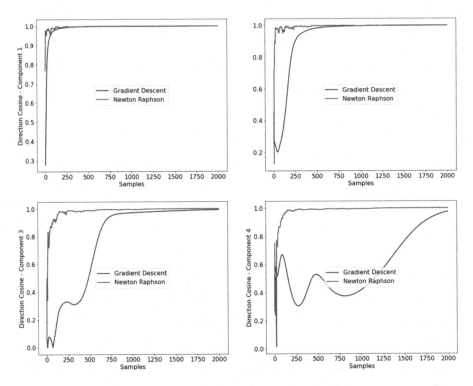

Figure 6-4. *Convergence of the first four principal eigenvectors of A by the gradient descent (6.4) and Newton-Raphson (6.12) algorithms for stationary data*

It is clear from Figures 6-2 through 6-4 that the steepest descent, conjugate direction, and Newton-Raphson algorithms converge faster than the gradient descent algorithm in spite of a careful selection of η_k for the gradient descent algorithm. Besides, the new algorithms do not require ad-hoc selections of η_k. Instead, the gain parameters α_k^i and β_k^i are computed from the online data sequence.

Comparison between the four algorithms show small differences between them for the first four principal eigenvectors of A. Among the three faster converging algorithms, the steepest descent algorithm (6.8) requires the smallest amount of computation per iteration. Therefore, these experiments show that the steepest descent adaptive algorithm (6.8) is most suitable for optimum speed and computation among the four algorithms presented here.

Experiments with Non-Stationary Data

In order to demonstrate the tracking ability of the algorithms with non-stationary data, I generated 500 samples of zero-mean 10-dimensional Gaussian data (i.e., $n=10$) with the covariance matrix stated before. I then abruptly changed the data sequence by generating 1,000 samples of zero-mean 10-dimensional Gaussian data with the covariance matrix below (the fifth covariance matrix from [Okada and Tomita 85] multiplied by 4):

$$
4 \begin{bmatrix}
0.0900 & 0.001 & -0.008 & -0.191 & -0.007 & 0.041 & -0.030 & -0.058 & 0.022 & 0.032 \\
0.001 & 0.092 & -0.011 & 0.008 & -0.014 & -0.020 & 0.012 & -0.010 & -0.021 & -0.002 \\
-0.008 & -0.011 & 0.082 & 0.082 & 0.014 & -0.020 & -0.058 & 0.105 & 0.004 & 0.023 \\
-0.191 & 0.008 & 0.082 & 5.680 & -0.096 & -0.015 & 0.646 & 0.219 & -0.238 & 0.218 \\
-0.007 & -0.014 & 0.014 & -0.096 & 0.076 & -0.035 & -0.040 & -0.023 & 0.027 & -0.014 \\
0.041 & -0.002 & -0.020 & -0.015 & -0.035 & 0.458 & 0.138 & -0.251 & 0.012 & 0.039 \\
-0.030 & 0.012 & -0.058 & 0.646 & -0.040 & 0.138 & 1.820 & -0.183 & -0.002 & 0.117 \\
-0.058 & -0.010 & 0.105 & 0.219 & -0.023 & -0.251 & -0.183 & 4.070 & -0.464 & 0.147 \\
0.022 & -0.021 & 0.004 & -0.238 & 0.027 & 0.012 & -0.002 & -0.464 & 0.263 & 0.054 \\
0.032 & -0.002 & 0.023 & 0.218 & -0.014 & 0.039 & 0.117 & 0.147 & 0.054 & 0.387
\end{bmatrix}.
$$

The eigenvalues of this covariance matrix are

23.3662, 16.5698, 6.8611, 1.8379, 1.5452, 0.7010, 0.3851, 0.3101, 0.2677, 0.2278,

which are drastically different from the previous eigenvalues. Figure 6-5 plots the 10-dimensional non-stationary data.

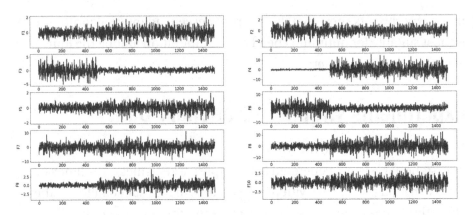

Figure 6-5. *10-dimensional non-stationary random data with abrupt changes after 500 samples*

I generated $\{A_k\}$ from $\{\mathbf{x}_k\}$ by using the algorithm (2.5 in Chapter 2) with $\beta=0.995$. I used the adaptive gradient descent (6.4), steepest descent (6.8), conjugate direction (6.11), and Newton-Raphson (6.12) algorithms on the random observation sequence $\{A_k\}$ and measured the convergence accuracy of the algorithms by computing the direction cosine at k^{th} update of each adaptive algorithm as shown in (6.18). I started all algorithms with $W_0 = 0.1*\text{ONE}$, where ONE is a 10 X 4 matrix whose all elements are ones. Here again I computed the first four eigenvectors (i.e., $p=4$).

Figures 6-6 through 6-8 show the iterates of the four algorithms to compute the first four principal eigenvectors of the two covariance matrices described before. For the conjugate direction method, I used the Hestenes-Stiefel [nonlinear conjugate gradient method, Wikipedia] method to compute β_k^i. For the steepest descent, conjugate direction, and Newton-Raphson methods, I chose α_k^i by solving a cubic equation as described in Sections 6.3 through 6.5.

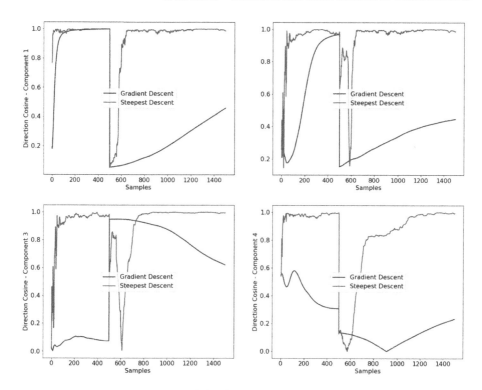

Figure 6-6. *Convergence of the first four principal eigenvectors of two covariance matrices by the gradient descent (6.4) and steepest descent (6.8) algorithms for non-stationary data*

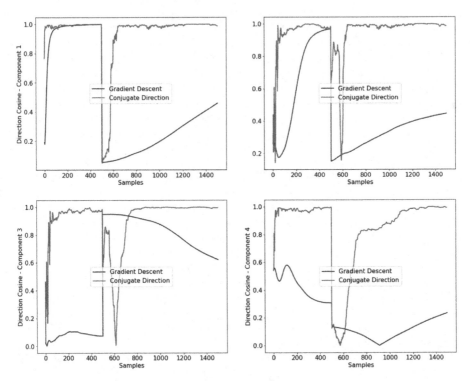

Figure 6-7. *Convergence of the first four principal eigenvectors of two covariance matrices by the gradient descent (6.4) and conjugate direction (6.11) algorithms for non-stationary data*

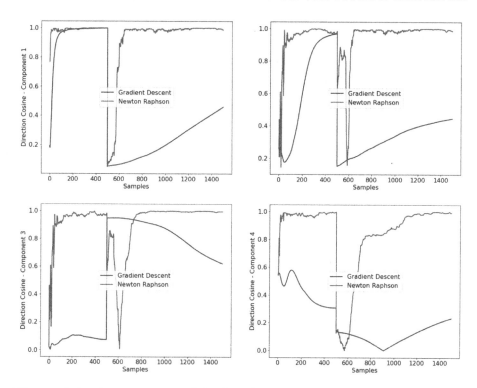

Figure 6-8. *Convergence of the first four principal eigenvectors of two covariance matrices by the gradient descent (6.4) and Newton-Raphson (6.12) algorithms for non-stationary data*

Once again, it is clear from Figures 6-6 through 6-8 that the steepest descent, conjugate direction, and Newton-Raphson algorithms converge faster and track the changes in data much better than the traditional gradient descent algorithm. In some cases, such as Figure 6-6 for the third principal eigenvector, the gradient descent algorithm fails as the data sequence changes, but the new algorithms perform correctly.

Comparison between the four algorithms in Figure 6-8 show small differences between them for the first four principal eigenvectors. Once again, among the three faster converging algorithms, since the steepest descent algorithm (6.8) requires the smallest amount of computation per iteration, it is most suitable for optimum speed and computation.

173

Comparison with State-of-the-Art Algorithms

I compared the steepest descent algorithm (6.8) with Yang's PASTd algorithm, Bannour and Sadjadi's RLS algorithm, and Fu and Dowling's CGET1 algorithm. I first tested the four algorithms on the stationary data described in Section 6.6.1.

I define ONE as a 10X4 matrix whose all elements are ones. The initial values for each algorithm are as follows:

1. The steepest descent algorithm:

    $W_0 = 0.1*$ONE.

2. Yang's PASTd algorithm:

    $W_0 = 0.1*$ONE, β=0.997 and $d_0^i = 0.2$ for i = 1, 2, ..., p $(p \leq n)$.

3. Bannour and Sadjadi's RLS algorithm:

    $W_0 = 0.1*$ONE and $P_0 =$ ONE.

4. Fu and Dowling's CGET1 algorithm:

    $W_0 = 0.1*$ONE and $A_0 = \mathbf{x}_k \mathbf{x}_k^T$.

I found that the performance of the PASTd and RLS algorithms depended considerably on the initial choices of d_0^i and P_0 respectively. I, therefore, chose the initial values that gave the best results for most experiments. The results of this experiment are shown in Figure 6-9.

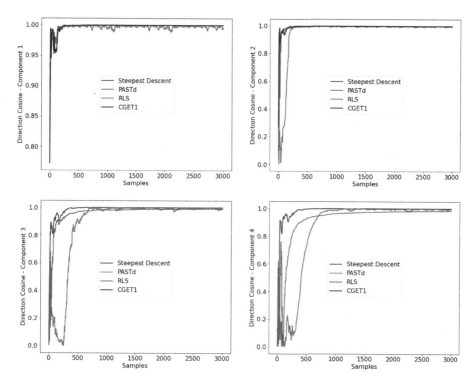

Figure 6-9. *Convergence of the first four principal eigenvectors of A by steepest descent (6.8), PASTd, RLS, and CGET1 algorithms for stationary data*

Observe from Figure 6-9 that the steepest descent and CGET1 algorithms perform quite well for all four principal eigenvectors. The RLS performed a little better than the PASTd algorithm for the minor eigenvectors. For the major eigenvectors, all algorithms performed well. The differences between the algorithms were evident for the minor (third and fourth) eigenvectors.

I next applied the four algorithms on non-stationary data described in Section 6.6.2 with β=0.995 in eq. (2.5, Chapter 2). The results of this experiment are shown in Figure 6-10.

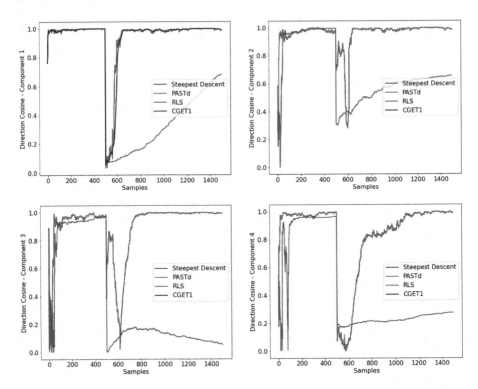

Figure 6-10. *Convergence of the first four principal eigenvectors of two covariance matrices by the steepest descent (6.8), PASTd, RLS, and CGET1 algorithms for non-stationary data*

Observe that the steepest descent and CGET1 algorithms perform quite well for all four principal eigenvectors. The PASTd algorithm performs better than the RLS algorithm in handling non-stationarity. This is expected since the PASTd algorithm accounts for non-stationarity with a forgetting factor of β=0.995, whereas the RLS algorithm has no such option.

6.7 Concluding Remarks

I presented an unconstrained objective function to obtain various new adaptive algorithms for PCA by using nonlinear optimization methods such as gradient descent, steepest descent, conjugate gradient, and Newton-Raphson. Comparison among these algorithms with stationary and non-stationary data show that the SD, CG, and NR algorithms have faster tracking abilities compared to the GD algorithm.

Further consideration should be given to the computational complexity of the algorithms. SD, CG, and NR algorithms have computational complexity of $O(pn^2)$. If, however, we use the estimate $A_k = \mathbf{x}_k \mathbf{x}_k^T$ instead of (6.4) in the GD algorithm, then the computational complexity drops to $O(pn)$, although the convergence gets slower. The CGET1 algorithm has complexity $O(pn^2)$. The PASTd and RLS algorithms have complexity $O(pn)$. However, their convergence is slower than the SD and CGET1 algorithms as shown in Figures 6-9 and 6-10. Further note that the GD algorithm can be implemented by parallel architecture as shown by the examples in [Cichocki and Unbehauen 93].

## CHAPTER 7

Generalized Eigenvectors

7.1 Introduction and Use Cases

This chapter is concerned with the adaptive solution of the generalized eigenvalue problems $A\Phi=B\Phi\Lambda$, $AB\Phi=\Phi\Lambda$, and $BA\Phi=\Phi\Lambda$, where A and B are real, symmetric, nXn matrices and B is positive definite. In particular, we shall consider the problem $A\Phi=B\Phi\Lambda$, although the remaining two problems are similar. The matrix pair (pencil) (A,B) is commonly referred to as a *symmetric-definite pencil* [Golub and VanLoan 83].

As seen before, the conventional (numerical analysis) method for evaluating Φ and Λ requires the computation of (A,B) after collecting all of the samples, and then the application of a numerical procedure [Golub and VanLoan 83]; in other words, the approach works in a *batch* fashion. In contrast, for the online case, matrices (A,B) are unknown. Instead, there are available two sequences of random matrices $\{A_k,B_k\}$ with $\lim_{k\to\infty}E[A_k]=A$ and $\lim_{k\to\infty}E[B_k]=B$. For every sample (A_k,B_k), we need to obtain the current estimates (Φ_k,Λ_k) of (Φ,Λ) respectively, such that (Φ_k,Λ_k) converge strongly to (Φ,Λ).

© Chanchal Chatterjee 2022
C. Chatterjee, *Adaptive Machine Learning Algorithms with Python*,
https://doi.org/10.1007/978-1-4842-8017-1_7

Application of GEVD in Pattern Recognition

In pattern recognition, there are problems where we are given samples ($x\in\Re^n$) from different populations or pattern classes. The well-known problem of linear discriminant analysis (LDA) [Chatterjee et al. Nov 97, May 97, Mar 97] seeks a transform $W\in\Re^{n\times p}$ ($p\leq n$), such that the interclass distance (measured by the scatter of the patterns around their mixture mean) is maximized, while at the same time the intra-class distance (measured by the scatter of the patterns around their respective class means) is as small as possible. The objective of this transform is to group the classes into well-separated clusters. The former scatter matrix, known as the mixture scatter matrix, is denoted by A, and the latter matrix, known as the within-class scatter matrix, is denoted by B [Fukunaga 90]. When the first column $\mathbf{w}$ of W is needed (i.e., $p=1$), the problem can be formulated in the constrained optimization framework as

$$\text{Maximize } \mathbf{w}^T A\mathbf{w} \text{ subject to } \mathbf{w}^T B\mathbf{w} = 1. \tag{7.1}$$

A twin problem to (7.1) is to maximize the Rayleigh quotient criterion [Golub and VanLoan 83] with respect to $\mathbf{w}$:

$$J(\mathbf{w};A,B) = \frac{\mathbf{w}^T A\mathbf{w}}{\mathbf{w}^T B\mathbf{w}}. \tag{7.2}$$

A solution to (7.1) or (7.2) leads to the generalized eigen-decomposition problem $A\mathbf{w}=\lambda B\mathbf{w}$, where λ is the largest generalized eigenvalue of A with respect to B. In general, the p columns of W are the $p\leq n$ orthogonal unit generalized eigenvectors $\boldsymbol{\phi}_1,...,\boldsymbol{\phi}_p$ of A with respect to B where

$$A\boldsymbol{\phi}_i = \lambda_i B\boldsymbol{\phi}_i, \ \boldsymbol{\phi}_i^T A\boldsymbol{\phi}_j = \lambda_i \delta_{ij}, \text{ and } \boldsymbol{\phi}_i^T B\boldsymbol{\phi}_j = \delta_{ij} \text{ for } i=1,...,p, \tag{7.3}$$

where $\lambda_1 > ... > \lambda_p > \lambda_{p+1} \geq ... \geq \lambda_n > 0$ are the p largest generalized eigenvalues of A with respect to B in descending order of magnitude. In summary, LDA is a powerful feature extraction tool for the class separability feature [Chatterjee May 97], and our adaptive algorithms are suited to this.

Application of GEVD in Signal Processing

Next, let's discuss an analogous problem of detecting a desired signal in the presence of interference. Here, we seek the optimum linear transform W for weighting the signal plus interference such that the desired signal is detected with maximum power and minimum interference. Given the matrix pair (A,B), where A is the correlation matrix of the signal plus interference plus noise and B is the correlation matrix of interference plus noise, we can formulate the signal detection problem as the constrained maximization problem in (7.1). Here, we maximize the signal power and minimize the power of the interference. The solution for W consists of the $p \leq n$ largest generalized eigenvectors of the matrix pencil (A,B). Adaptive generalized eigen-decomposition algorithms also allow the tracking of slow changes in the incoming data [Chatterjee et al. Nov 97, Mar 97; Chen et al. 2000].

Methods for Generalized Eigen-Decomposition

We first define the problem for the non-adaptive case. Each of the three generalized eigenvalue problems ($A\Phi = B\Phi\Lambda$, $AB\Phi = \Phi\Lambda$, and $BA\Phi = \Phi\Lambda$, where A and B are real, symmetric, nXn matrices and B is positive definite) can be reduced to a standard symmetric eigenvalue problem using a Cholesky factorization of B as either $B = LL^T$ or $B = U^TU$. With $B = LL^T$, we can write $A\Phi = B\Phi\Lambda$ as

$$(L^{-1}AL^{-T})(L^T\Phi) = (L^T\Phi)\Lambda \text{ or } C\Psi = \Psi\Lambda. \tag{7.4}$$

Here C is the symmetric matrix $C = L^{-1} A L^{-T}$ and $\Psi = L^T \Phi$. Table 7-1 summarizes how each of the three types of problems can be reduced to the standard form $C\Psi = \Psi\Lambda$, and how the eigenvectors Φ of the original problem may be recovered from the eigenvectors Ψ of the reduced problem.

Table 7-1. *Types of Generalized Eigen-Decomposition problems and their solutions*

Type of Problem	Factorization of B	Reduction	Generalized Eigenvectors
$A\Phi = B\Phi\Lambda$	$B = LL^T$	$C = L^{-1} A L^{-T}$	$\Phi = L^{-T}\Psi$
	$B = U^T U$	$C = U^{-T} A U^{-1}$	$\Phi = U^{-1}\Psi$
$AB\Phi = \Phi\Lambda$	$B = LL^T$	$C = L^T A L$	$\Phi = L^{-T}\Psi$
	$B = U^T U$	$C = U A U^T$	$\Phi = U^{-1}\Psi$
$BA\Phi = \Phi\Lambda$	$B = LL^T$	$C = L^T A L$	$\Phi = L\Psi$
	$B = U^T U$	$C = U A U^T$	$\Phi = U^T\Psi$

In the adaptive case, we can extend these techniques by first (adaptively) computing a matrix W_k for each sample B_k, where W_k tends to the inverse Cholesky factorization L^{-1} of B with probability one (w.p.1) as $k \to \infty$. Any of the algorithms in Chapter 3 can be considered here. Next, we consider a sequence $\{C_k = W_{k-1} A_k W_{k-1}^T\}$, which is used to adaptively compute a matrix V_k, where V_k tends to the eigenvector matrix of $\lim_{k \to \infty} E[C_k]$ w.p.1 as $k \to \infty$. Any of the algorithms in Chapters 5 and 6 can be considered for this purpose. In conjunction, the two steps yield $W_k V_k$, which is proven to converge w.p.1 to Φ as $k \to \infty$. Thus, the two steps can proceed *simultaneously* and converge strongly to the eigenvector matrix Φ. A full description of this method is given in [Chatterjee et al. May 97, Mar 97]. In this chapter, I offer a variety of new techniques to solve the generalized eigen-decomposition problem for the adaptive case.

Outline of This Chapter

In Section 7.2, I list the objective functions from which I derive the adaptive generalized eigen-decomposition algorithms. In Sections 7.3, 7.4, and 7.5, I present adaptive algorithms for the homogeneous, deflation, and weighted variations, respectively, from the OJA objective function.

In Sections 7.6, 7.7, and 7.8, I analyze the same three variations for the mean squared error (XU) objective function and convergence proofs for the deflation case. In Sections 7.9, 7.10, and 7.11, I discuss algorithms derived from the penalty function (PF) objective function. In Sections 7.12, 7.13, and 7.14, I consider the augmented Lagrangian 1 (AL1) objective function, and in Sections 7.15, 7.16, and 7.17, I present the augmented Lagrangian 2 (AL2) objective function. In Sections 7.18, 7.19, and 7.20, I present the information theory (IT) criterion, and in Sections 7.21, 7.22, and 7.23, I describe the Rayleigh quotient (RQ) criterion. In Section 7.24, I discusses the experimental results, and in Section 7.25, I present conclusions.

7.2 Algorithms and Objective Functions

Similar to the PCA algorithms (Chapter 5), in this chapter, I present several adaptive algorithms for generalized eigenvector computation. I consider two asymptotically stationary sequences $\{\mathbf{x}_k \in \Re^n\}$ and $\{\mathbf{y}_k \in \Re^n\}$ that have been centered to zero mean. We can represent the corresponding online correlation matrices $\{A_k, B_k\}$ of $\{\mathbf{x}_k, \mathbf{y}_k\}$ either by their instantaneous values $\{\mathbf{x}_k \mathbf{x}_k^T, \mathbf{y}_k \mathbf{y}_k^T\}$ or by their running averages by (2.3). If, however, $\{\mathbf{x}_k, \mathbf{y}_k\}$ are non-stationary, we can construct correlation matrices $\{A_k, B_k\}$ out of data samples $\{\mathbf{x}_k, \mathbf{y}_k\}$ by (2.5).

Summary of Objective Functions for Adaptive GEVD Algorithms

Conforming to the methodology in Section 2.3, for each algorithm, I describe objective functions and derive the adaptive algorithms for them. The objective functions are

- Oja's objective function (OJA),

- Xu's mean squared error objective function (XU),

- Penalty function method (PF),

- Augmented Lagrangian Method 1 (AL1),

- Augmented Lagrangian Method 2 (AL2),

- Information theory criterion (IT), and

- Rayleigh quotient criterion (RQ).

Same as the PCA case, there are three variations of algorithms derived from each objective function. They are

1. *Homogeneous Adaptive Rule:* These algorithms do not compute the true normalized generalized eigenvectors with decreasing eigenvalues.

2. *Deflation Adaptive Rule:* Here, we produce unit generalized eigenvectors with decreasing eigenvalues. However, the training is sequential, thereby making the training process harder for parallel implementations.

3. *Weighted Adaptive Rule:* These algorithms are obtained by using a different scalar weight for each generalized eigenvector, making them normalized and in the order of decreasing eigenvalues.

Summary of Generalized Eigenvector Algorithms

For all algorithms, I describe an *objective function* $J(\mathbf{w}_i; A, B)$ and an update rule of the form

$$W_{k+1} = W_k + \eta_k h\left(W_k, A_k, B_k\right),$$

where $h(W_k, A_k, B_k)$ follows certain continuity and regularity properties [Ljung 77,92] and are given in Table 7-2.

Table 7-2. *List of Adaptive Generalized Eigen-Decomposition Algorithms*

Alg.	Type	Adaptive Algorithm $h(W_k, A_k)$
OJA	Homogeneous	$A_k W_k - B_k W_k W_k^T A_k W_k$
	Deflation	$A_k W_k - B_k W_k \mathrm{UT}\left(W_k^T A_k W_k\right)$
	Weighted	$A_k W_k C - B_k W_k C W_k^T A_k W_k$
XU	Homogeneous	$2 A_k W_k - A_k W_k W_k^T B_k W_k - B_k W_k W_k^T A_k W_k$
	Deflation	$2 A_k W_k - A_k W_k \mathrm{UT}\left(W_k^T B_k W_k\right) - B_k W_k \mathrm{UT}\left(W_k^T A_k W_k\right)$
	Weighted	$2 A_k W_k C - B_k W_k C W_k^T A_k W_k - A_k W_k C W_k^T B_k W_k$
PF	Homogeneous	$A_k W_k - \mu B_k W_k \left(W_k^T B_k W_k - I_p\right)$
	Deflation	$A_k W_k - \mu B_k W_k \mathrm{UT}\left(W_k^T B_k W_k - I_p\right)$
	Weighted	$A_k W_k C - \mu B_k W_k C \left(W_k^T B_k W_k - I_p\right)$
AL1	Homogeneous	$A_k W_k - B_k W_k W_k^T A_k W_k - \mu B_k W_k \left(W_k^T B_k W_k - I_p\right)$
	Deflation	$A_k W_k - B_k W_k \mathrm{UT}\left(W_k^T A_k W_k\right) - \mu B_k W_k \mathrm{UT}\left(W_k^T B_k W_k - I_p\right)$
	Weighted	$A_k W_k C - B_k W_k C W_k^T A_k W_k - \mu B_k W_k C \left(W_k^T B_k W_k - I_p\right)$

(continued)

Table 7-2. (*continued*)

Alg.	Type	Adaptive Algorithm $h(W_k, A_k)$
AL2	Homogeneous	$2A_k W_k - B_k W_k W_k^T A_k W_k - A_k W_k W_k^T B_k W_k - \mu B_k W_k \left(W_k^T B_k W_k - I_p \right)$
	Deflation	$2A_k W_k - B_k W_k \mathrm{UT}\left(W_k^T A_k W_k \right) - A_k W_k \mathrm{UT}\left(W_k^T B_k W_k \right) - \mu B_k W_k \mathrm{UT}\left(W_k^T B_k W_k - I_p \right)$
	Weighted	$2A_k W_k C - B_k W_k C W_k^T A_k W_k - A_k W_k C W_k^T B_k W_k - \mu B_k W_k C \left(W_k^T B_k W_k - I_p \right)$
IT	Homogeneous	$\left(A_k W_k - B_k W_k W_k^T A_k W_k \right)\ \mathrm{DIAG}\left(W_k^T A_k W_k \right)^{-1}$
	Deflation	$\left(A_k W_k - B_k W_k \mathrm{UT}\left(W_k^T A_k W_k \right) \right)\ \mathrm{DIAG}\left(W_k^T A_k W_k \right)^{-1}$
	Weighted	$\left(A_k W_k C - B_k W_k C W_k^T A_k W_k \right)\ \mathrm{DIAG}\left(W_k^T A_k W_k \right)^{-1}$
RQ	Homogeneous	$\left(A_k W_k - B_k W_k W_k^T A_k W_k \right)\ \mathrm{DIAG}\left(W_k^T B_k W_k \right)^{-1}$
	Deflation	$\left(A_k W_k - B_k W_k \mathrm{UT}\left(W_k^T A_k W_k \right) \right)\ \mathrm{DIAG}\left(W_k^T B_k W_k \right)^{-1}$
	Weighted	$\left(A_k W_k C - B_k W_k C W_k^T A_k W_k \right)\ \mathrm{DIAG}\left(W_k^T B_k W_k \right)^{-1}$

In the following discussions, I denote $\Phi = [\phi_1 \ \dots \ \phi_n] \in \mathfrak{R}^{n \times n}$ as the orthonormal generalized eigenvector matrix of A with respect to B, and $\Lambda = diag(\lambda_1, \dots, \lambda_n)$ as the generalized eigenvalue matrix, such that $\lambda_1 > \lambda_2 > \dots > \lambda_p > \lambda_{p+1} \geq \dots \geq \lambda_n > 0$. I use the subscript (i) to denote the ith permutation of the indices $\{1, 2, \dots, n\}$.

7.3 OJA GEVD Algorithms

OJA Homogeneous Algorithm

The objective function for the OJA homogeneous algorithm can be written as

$$J\left(\mathbf{w}_k^i;A_k,B_k\right)=-\mathbf{w}_k^{i^T}A_kB_k^{-1}A_k\mathbf{w}_k^i+\frac{1}{2}\left(\mathbf{w}_k^{i^T}A_k\mathbf{w}_k^i\right)^2+\sum_{j=1,j\neq i}^{p}\left(\mathbf{w}_k^{i^T}A_k\mathbf{w}_k^j\right)^2,\quad(7.5)$$

for $i=1,...,p$ $(p\leq n)$. From the gradient of (7.5) with respect to $\mathbf{w}_k^i$ we obtain the following adaptive algorithm:

$$\mathbf{w}_{k+1}^i=\mathbf{w}_k^i-\eta_kB_kA_k^{-1}\nabla_{\mathbf{w}_k^i}J\left(\mathbf{w}_k^i;A_k,B_k\right)\text{ for }i=1,...,p,\quad(7.6)$$

where η_k is a decreasing gain constant. Defining $W_k=\left[\mathbf{w}_k^1...\mathbf{w}_k^p\right]$, from (7.6) we get

$$W_{k+1}=W_k+\eta_k\left(A_kW_k-B_kW_kW_k^TA_kW_k\right).\quad(7.7)$$

OJA Deflation Algorithm

The objective function for the OJA deflation adaptive GEVD algorithm is

$$J\left(\mathbf{w}_k^i;A_k,B_k\right)=-\mathbf{w}_k^{i^T}A_kB_k^{-1}A_k\mathbf{w}_k^i+\frac{1}{2}\left(\mathbf{w}_k^{i^T}A_k\mathbf{w}_k^i\right)^2+\sum_{j=1}^{i-1}\left(\mathbf{w}_k^{i^T}A_k\mathbf{w}_k^j\right)^2,\quad(7.8)$$

for $i=1,...,p$. From the gradient of (7.8) with respect to $\mathbf{w}_k^i$, we obtain the OJA deflation adaptive gradient descent algorithm as

$$\mathbf{w}_{k+1}^i=\mathbf{w}_k^i-\eta_k\left(A_k\mathbf{w}_k^i-\sum_{j=1}^{i}B_k\mathbf{w}_k^j\mathbf{w}_k^{j^T}A_k\mathbf{w}_k^i\right),\quad(7.9)$$

for $i=1,...,p$ $(p\leq n)$. The matrix form of the algorithm is

$$W_{k+1}=W_k+\eta_k\left(A_kW_k-B_kW_k\text{UT}\left(W_k^TA_kW_k\right)\right),\quad(7.10)$$

where UT[·] sets all elements below the diagonal of its matrix argument to zero.

187

OJA Weighted Algorithm

The objective function for the OJA weighted adaptive GEVD algorithm is

$$J\left(\mathbf{w}_k^i; A_k, B_k\right) = -c_i {\mathbf{w}_k^i}^T A_k B_k^{-1} A_k \mathbf{w}_k^i + \frac{c_i}{2}\left({\mathbf{w}_k^i}^T A_k \mathbf{w}_k^i\right)^2 +$$

$$\sum_{j=1, j\neq i}^{p} c_j \left({\mathbf{w}_k^i}^T A_k \mathbf{w}_k^j\right)^2 \tag{7.11}$$

for $i=1,\ldots,p$, where $c_1,\ldots,c_p$ ($p\leq n$) are small positive numbers satisfying

$$c_1 > c_2 > \ldots > c_p > 0,\ p \leq n. \tag{7.12}$$

Given a diagonal matrix $C = diag\left(c_1, \ldots, c_p\right)$, $p \leq n$, the OJA weighted adaptive algorithm is

$$W_{k+1} = W_k + \eta_k \left(A_k W_k C - B_k W_k C W_k^T A_k W_k\right). \tag{7.13}$$

OJA Algorithm Python Code

The following Python code works with multidimensional data
X[nDim,nSamples] and Y[nDim,nSamples]:

```python
from numpy import linalg as la
A = np.zeros(shape=(nDim,nDim)) # stores adaptive
                                  correlation matrix
B = np.zeros(shape=(nDim,nDim)) # stores adaptive
                                  correlation matrix
W2 = 0.1 * np.ones(shape=(nDim,nEA)) # weight vectors of all
                                       algorithms

W3 = W2
c = [2.6-0.3*k for k in range(nEA)]
C = np.diag(c)
I = np.identity(nDim)
```

```
for epoch in range(nEpochs):
    for iter in range(nSamples):
        cnt = nSamples*epoch + iter
        # Update data correlation matrices A,B with current
        data vectors x,y
        x = X[:,iter]
        x = x.reshape(nDim,1)
        A = A + (1.0/(1 + cnt))*((np.dot(x, x.T)) - A)
        y = Y[:,iter]
        y = y.reshape(nDim,1)
        B = B + (1.0/(1 + cnt))*((np.dot(y, y.T)) - B)
        # Deflated Gradient Descent
        W2 = W2 + (1/(150 + cnt))*(A @ W2 - B @ W2
                                   @ np.triu(W2.T @ A @ W2))
        # Weighted Gradient Descent
        W3 = W3 + (1/(500 + cnt))*(A @ W3 @ C - B @ W3 @ C
                                   @ (W3.T @ A @ W3))
```

7.4 XU GEVD Algorithms

XU Homogeneous Algorithm

The objective function for the XU homogeneous adaptive GEVD algorithm is

$$
J\left(\mathbf{w}_k^i; A_k, B_k\right) = -2\mathbf{w}_k^{i^T} A_k \mathbf{w}_k^i + \left(\mathbf{w}_k^{i^T} A_k \mathbf{w}_k^i\right)\left(\mathbf{w}_k^{i^T} B_k \mathbf{w}_k^i\right) +
$$
$$
2\sum_{j=1, j \neq i}^{p} \mathbf{w}_k^{i^T} A_k \mathbf{w}_k^j \mathbf{w}_k^{j^T} B_k \mathbf{w}_k^i, \tag{7.14}
$$

for $i=1,...,p$ $(p{\leq}n)$. From the gradient of (7.14) with respect to $\mathbf{w}_k^i$, we obtain the XU homogeneous adaptive gradient descent algorithm as

$$\mathbf{w}_{k+1}^i = \mathbf{w}_k^i + \eta_k \left(2A_k\mathbf{w}_k^i - \sum_{j=1}^{p} A_k\mathbf{w}_k^j\mathbf{w}_k^{j^T}B_k\mathbf{w}_k^i - \sum_{j=1}^{p} B_k\mathbf{w}_k^j\mathbf{w}_k^{j^T}A_k\mathbf{w}_k^i \right) \quad (7.15)$$

for $i=1,...,p$, whose matrix form is

$$W_{k+1} = W_k + \eta_k \left(2A_kW_k - A_kW_kW_k^TB_kW_k - B_kW_kW_k^TA_kW_k \right). \quad (7.16)$$

XU Deflation Algorithm

The objective function for the XU deflation adaptive GEVD algorithm is

$$J\left(\mathbf{w}_k^i;A_k,B_k\right) = -2\mathbf{w}_k^{i^T}A_k\mathbf{w}_k^i + \left(\mathbf{w}_k^{i^T}A_k\mathbf{w}_k^i\right)\left(\mathbf{w}_k^{i^T}B_k\mathbf{w}_k^i\right) +$$

$$2\sum_{j=1}^{i-1}\mathbf{w}_k^{i^T}A_k\mathbf{w}_k^j\mathbf{w}_k^{j^T}B_k\mathbf{w}_k^i, \quad (7.17)$$

for $i=1,...,p$ $(p{\leq}n)$. From the gradient of (7.17) with respect to $\mathbf{w}_k^i$, we obtain

$$W_{k+1} = W_k + \eta_k\left(2A_kW_k - A_kW_k\mathrm{UT}\left(W_k^TB_kW_k\right) - B_kW_k\mathrm{UT}\left(W_k^TA_kW_k\right)\right), \quad (7.18)$$

where UT[·] sets all elements below the diagonal of its matrix argument to zero. Chatterjee et al. [Mar 00, Thms 1, 2] proved that W_k converges with probability one to $[\pm\boldsymbol{\phi}_1 \pm\boldsymbol{\phi}_2 ... \pm\boldsymbol{\phi}_p]$ as $k{\rightarrow}\infty$.

XI Weighted Algorithm

The objective function for the XU weighted adaptive GEVD algorithm is

$$J\left(\mathbf{w}_k^i;A_k,B_k\right) = -2c_i\mathbf{w}_k^{i^T}A_k\mathbf{w}_k^i + c_i\left(\mathbf{w}_k^{i^T}A_k\mathbf{w}_k^i\right)\left(\mathbf{w}_k^{i^T}B_k\mathbf{w}_k^i\right) +$$

$$2\sum_{j=1,j{\neq}i}^{p}c_j\mathbf{w}_k^{i^T}A_k\mathbf{w}_k^j\mathbf{w}_k^{j^T}B_k\mathbf{w}_k^i \quad (7.19)$$

for $i=1,\ldots,p$ $(p \leq n)$, where $c_1,\ldots,c_p$ are small positive numbers satisfying (7.12). The adaptive algorithm is

$$W_{k+1} = W_k + \eta_k \left(2A_kW_kC - B_kW_kCW_k^T A_kW_k - A_kW_kCW_k^T B_kW_k \right), \qquad (7.20)$$

where $C = diag(c_1,\ldots,c_p)$.

XU Algorithm Python Code

The following Python code works with multidimensional data X[nDim,nSamples] and Y[nDim,nSamples]:

```python
from numpy import linalg as la
A   np.zeros(shape=(nDim,nDim)) # stores adaptive
                                correlation matrix
B   = np.zeros(shape=(nDim,nDim)) # stores adaptive
                                correlation matrix
W2 = 0.1 * np.ones(shape=(nDim,nEA)) # weight vectors of all
                                algorithms
W3 = W2
c = [2.6-0.3*k for k in range(nEA)]
C = np.diag(c)
for epoch in range(nEpochs):
    for iter in range(nSamples):
        cnt = nSamples*epoch + iter
        # Update data correlation matrices A,B with current
        data vectors x,y
        x = X[:,iter]
        x = x.reshape(nDim,1)
        A = A + (1.0/(1 + cnt))*((np.dot(x, x.T)) - A)
        y = Y[:,iter]
        y = y.reshape(nDim,1)
        B = B + (1.0/(1 + cnt))*((np.dot(y, y.T)) - B)
```

```
# Deflated Gradient Descent
W2 = W2 + (1/(100 + cnt))*(A @ W2 - 0.5 * B @ W2 @ \
                          np.triu(W2.T @ A @ W2)- 0.5*
                          A@ W2 @ np.triu(W2.T@ B @ W2))
# Weighted Gradient Descent
W3 = W3 + (1/(300 + cnt))*(A @ W3 @ C - 0.5 * B @ W3 @
                          C @ \(W3.T @ A @ W3) - 0.5 *
                          A @ W3 @ C @ (W3.T @
                          B @ W3))
```

7.5 PF GEVD Algorithms

PF Homogeneous Algorithm

We obtain the objective function for the PF homogeneous generalized eigenvector algorithm by writing the Rayleigh quotient criterion (7.2) as the following penalty function:

$$J\left(\mathbf{w}_k^i; A_k, B_k\right) = -\mathbf{w}_k^{i^T} A_k \mathbf{w}_k^i + \mu\left(\sum_{j=1, j\neq i}^{p}\left(\mathbf{w}_k^{j^T} B_k \mathbf{w}_k^i\right)^2 + \frac{1}{2}\left(\mathbf{w}_k^{i^T} B_k \mathbf{w}_k^i - 1\right)^2\right),\quad(7.21)$$

where $\mu > 0$ and $i=1,\ldots,p$ $(p\leq n)$. From the gradient of (7.21) with respect to $\mathbf{w}_k^i$, we obtain the PF homogeneous adaptive algorithm:

$$W_{k+1} = W_k + \eta_k\left(A_k W_k - \mu B_k W_k\left(W_k^T B_k W_k - I_p\right)\right),\quad(7.22)$$

where I_p is a $p\mathsf{X}p$ identity matrix.

PF Deflation Algorithm

The objective function for the PF deflation GEVD algorithm is

$$J\left(\mathbf{w}_k^i; A_k, B_k\right) = -\mathbf{w}_k^{i^T} A_k \mathbf{w}_k^i + \mu\left(\sum_{j=1}^{i-1}\left(\mathbf{w}_k^{j^T} B_k \mathbf{w}_k^i\right)^2 + \frac{1}{2}\left(\mathbf{w}_k^{i^T} B_k \mathbf{w}_k^i - 1\right)^2\right),\quad(7.23)$$

where $\mu > 0$ and $i=1,...,p$. The adaptive algorithm is

$$W_{k+1} = W_k + \eta_k \left(A_k W_k - \mu B_k W_k UT \left(W_k^T B_k W_k - I_p \right) \right), \tag{7.24}$$

where UT[·] sets all elements below the diagonal of its matrix argument to zero.

PF Weighted Algorithm

The objective function for the PF weighted GEVD algorithm is

$$J\left(\mathbf{w}_k^i; A_k, B_k \right) = -c_i {\mathbf{w}_k^i}^T A_k \mathbf{w}_k^i + \mu \left(\sum_{j=1, j \neq i}^{p} c_i \left({\mathbf{w}_k^j}^T B_k \mathbf{w}_k^i \right)^2 + \frac{c_j}{2} \left({\mathbf{w}_k^i}^T B_k \mathbf{w}_k^i - 1 \right)^2 \right), \tag{7.25}$$

where $c_1 > c_2 > ... > c_p > 0$ $(p \leq n)$, $\mu > 0$, and $i=1,...,p$. The corresponding adaptive algorithm is

$$W_{k+1} = W_k + \eta_k \left(A_k W_k C - \mu B_k W_k C \left(W_k^T B_k W_k - I_p \right) \right), \tag{7.26}$$

where $C = diag\left(c_1, ..., c_p \right)$.

PF Algorithm Python Code

The following Python code works with multidimensional data X[nDim,nSamples] and Y[nDim,nSamples]:

```
from numpy import linalg as la
A  = np.zeros(shape=(nDim,nDim)) # stores adaptive
                                correlation matrix
B  = np.zeros(shape=(nDim,nDim)) # stores adaptive
                                correlation matrix
W2 = 0.1 * np.ones(shape=(nDim,nEA)) # weight vectors of all
                                algorithms
W3 = W2
```

```
c = [2.6-0.3*k for k in range(nEA)]
C = np.diag(c)
I = np.identity(nDim)
for epoch in range(nEpochs):
    for iter in range(nSamples):
        cnt = nSamples*epoch + iter
        # Update data correlation matrices A,B with current
        data vectors x,y
        x = X[:,iter]
        x = x.reshape(nDim,1)
        A = A + (1.0/(1 + cnt))*((np.dot(x, x.T)) - A)
        y = Y[:,iter]
        y = y.reshape(nDim,1)
        B = B + (1.0/(1 + cnt))*((np.dot(y, y.T)) - B)
        # Deflated Gradient Descent
        W2 = W2 + (1/(100 + cnt))*(A @ W2 - mu * B @ W2 @ \
                                  np.triu((W2.T @ B @ W2) - I))
        # Weighted Gradient Descent
        W3 = W3 + (1/(500 + cnt))*(A @ W3 @ C - mu *
                                  B @ W3 @ C @ \
                                  ((W3.T @ B @ W3) - I))
```

7.6 AL1 GEVD Algorithms

AL1 Homogeneous Algorithm

We apply the augmented Lagrangian method of nonlinear optimization to
the Rayleigh quotient criterion (7.4) to obtain the objective function for the
AL1 homogeneous GEVD algorithm as

$$J\left(\mathbf{w}_k^i; A_k, B_k\right) = -\mathbf{w}_k^{i^T} A_k \mathbf{w}_k^i + \alpha\left(\mathbf{w}_k^{i^T} B_k \mathbf{w}_k^i - 1\right) + 2\sum_{j=1, j \neq i}^{p} \beta_j \mathbf{w}_k^{j^T} B_k \mathbf{w}_k^i,$$

$$+\mu\left(\sum_{j=1, j \neq i}^{p}\left(\mathbf{w}_k^{j^T} B_k \mathbf{w}_k^i\right)^2 + \frac{1}{2}\left(\mathbf{w}_k^{i^T} B_k \mathbf{w}_k^i - 1\right)^2\right), \tag{7.27}$$

for $i=1,\dots,p$ $(p \leq n)$, where $(\alpha, \beta_1, \beta_2, \dots, \beta_p)$ are Lagrange multipliers and μ is a positive penalty constant. Taking the gradient of $J\left(\mathbf{w}_k^i; A_k, B_k\right)$ with respect to $\mathbf{w}_k^i$ and equating the gradient to $\mathbf{0}$ and using the constraint $\mathbf{w}_k^{j^T} B_k \mathbf{w}_k^i = \delta_{ij}$, we obtain

$$\alpha = \mathbf{w}_k^{i^T} A_k \mathbf{w}_k^i \text{ and } \beta_j = \mathbf{w}_k^{j^T} A_k \mathbf{w}_k^i \text{ for } j=1,\dots,p. \tag{7.28}$$

Replacing $(\alpha, \beta_1, \beta_2, \dots, \beta_p)$ in the gradient of (7.27), we obtain the AL1 homogeneous adaptive gradient descent generalized eigenvector algorithm:

$$\mathbf{w}_{k+1}^i = \mathbf{w}_k^i + \eta_k\left(A_k \mathbf{w}_k^i - \sum_{j=1}^{p} B_k \mathbf{w}_k^j\left(\mathbf{w}_k^{j^T} A_k \mathbf{w}_k^i\right) - \mu \sum_{j=1}^{p} B_k \mathbf{w}_k^j\left(\mathbf{w}_k^{j^T} B_k \mathbf{w}_k^i - \delta_{ij}\right)\right) \tag{7.29}$$

where $\mu > 0$. Defining $W_k = \left[\mathbf{w}_k^1 \dots \mathbf{w}_k^p\right]$, we obtain

$$W_{k+1} = W_k + \eta_k\left(\left(A_k W_k - B_k W_k W_k^T A_k W_k - \mu B_k W_k\left(W_k^T B_k W_k - I_p\right)\right)\right), \tag{7.30}$$

where I_p is a $p \times p$ identity matrix.

AL1 Deflation Algorithm

The objective function for the AL1 deflation GEVD algorithm is

$$J\left(\mathbf{w}_k^i; A_k, B_k\right) = -\mathbf{w}_k^{i^T} A_k \mathbf{w}_k^i + \alpha\left(\mathbf{w}_k^{i^T} B_k \mathbf{w}_k^i - 1\right) + 2\sum_{j=1}^{i-1} \beta_j \mathbf{w}_k^{j^T} B_k \mathbf{w}_k^i$$

$$+\mu\left(\sum_{j=1}^{i-1}\left(\mathbf{w}_k^{j^T} B_k \mathbf{w}_k^i\right)^2 + \frac{1}{2}\left(\mathbf{w}_k^{i^T} B_k \mathbf{w}_k^i - 1\right)^2\right), \tag{7.31}$$

for $i=1,...,p$ ($p\leq n$). Following the steps in Section 7.6.1 we obtain the adaptive algorithm:

$$W_{k+1} = W_k + \eta_k \left(\left(A_k W_k - B_k W_k \mathrm{UT}\left(W_k^T A_k W_k \right) - \mu B_k W_k \mathrm{UT}\left(W_k^T B_k W_k - I_p \right) \right) \right). \quad (7.32)$$

AL1 Weighted Algorithm

The objective function for the AL1 weighted GEVD algorithm is

$$J\left(\mathbf{w}_k^i ; A_k, B_k \right) = -c_i \mathbf{w}_k^{i^T} A_k \mathbf{w}_k^i + \alpha c_i \left(\mathbf{w}_k^{i^T} B_k \mathbf{w}_k^i - 1 \right) + 2 \sum_{j=1, j\neq i}^{p} \beta_j c_j \mathbf{w}_k^{j^T} B_k \mathbf{w}_k^i ,$$

$$+ \mu \left(\sum_{j=1, j\neq i}^{p} c_j \left(\mathbf{w}_k^{j^T} B_k \mathbf{w}_k^i \right)^2 + \frac{c_i}{2} \left(\mathbf{w}_k^{i^T} B_k \mathbf{w}_k^i - 1 \right)^2 \right), \quad (7.33)$$

for $i=1,...,p$ ($p\leq n$), where $(\alpha, \beta_1, \beta_2, ..., \beta_p)$ are Lagrange multipliers, μ is a positive penalty constant, and $c_1 > c_2 > ... > c_p > 0$. The adaptive algorithm is

$$W_{k+1} = W_k + \eta_k \left(\left(A_k W_k C - B_k W_k C W_k^T A_k W_k - \mu B_k W_k C \left(W_k^T B_k W_k - I_p \right) \right) \right), \quad (7.34)$$

where $C = diag\left(c_1, ..., c_p \right)$.

AL1 Algorithm Python Code

The following Python code works with multidimensional data X[nDim,nSamples] and Y[nDim,nSamples]:

```
from numpy import linalg as la
A  = np.zeros(shape=(nDim,nDim)) # stores adaptive
                                   correlation matrix
B  = np.zeros(shape=(nDim,nDim)) # stores adaptive
                                   correlation matrix
```

```python
W2 = 0.1 * np.ones(shape=(nDim,nEA)) # weight vectors of all
                                     algorithms
W3 = W2
c = [2.6-0.3*k for k in range(nEA)]
C = np.diag(c)
I = np.identity(nDim)
mu = 2
for epoch in range(nEpochs):
    for iter in range(nSamples):
        cnt = nSamples*epoch + iter
        # Update data correlation matrices A,B with current
        data vectors x,y
        x = X[:,iter]
        x = x.reshape(nDim,1)
        A = A + (1.0/(1 + cnt))*((np.dot(x, x.T)) - A)
        y = Y[:,iter]
        y = y.reshape(nDim,1)
        B = B + (1.0/(1 + cnt))*((np.dot(y, y.T)) - B)
        # Deflated Gradient Descent
        W2 = W2 + (1/(500 + cnt))*(A @ W2 - B @ W2 @
                    np.triu(W2.T @ A @ W2) \
                    - mu * B @ W2 @ np.triu((W2.T @ B @ W2) - I))
        # Weighted Gradient Descent
        W3 = W3 + (1/(1000 + cnt))*(A @ W3 @ C - B @ W3 @ C @
                    (W3.T @ A @ W3) \
                    - mu * B @ W3 @ C @ ((W3.T @ B @ W3) - I))
```

7.7 AL2 GEVD Algorithms

AL2 Homogeneous Algorithm

The unconstrained objective function for the AL2 homogeneous GEVD algorithm is

$$J\left(\mathbf{w}_k^i; A_k, B_k\right) = -2\mathbf{w}_k^{i^T} A_k \mathbf{w}_k^i + \left(\mathbf{w}_k^{i^T} A_k \mathbf{w}_k^i\right)\left(\mathbf{w}_k^{i^T} B_k \mathbf{w}_k^i\right) +$$

$$2\sum_{j=1, j \neq i}^{p} \mathbf{w}_k^{i^T} A_k \mathbf{w}_k^j \mathbf{w}_k^{j^T} B_k \mathbf{w}_k^i + \mu\left(\sum_{j=1, j \neq i}^{p}\left(\mathbf{w}_k^{j^T} B_k \mathbf{w}_k^i\right)^2 + \frac{1}{2}\left(\mathbf{w}_k^{i^T} B_k \mathbf{w}_k^i - 1\right)^2\right), \quad (7.35)$$

for $i=1,\ldots,p$, where μ is a positive penalty constant. From (7.35), we obtain the adaptive gradient descent algorithm:

$$\mathbf{w}_{k+1}^i = \mathbf{w}_k^i + \eta_k \left(\begin{array}{c} 2A_k \mathbf{w}_k^i - \sum_{j=1}^{p} B_k \mathbf{w}_k^j \mathbf{w}_k^{j^T} A_k \mathbf{w}_k^i - \sum_{j=1}^{p} A_k \mathbf{w}_k^j \mathbf{w}_k^{j^T} B_k \mathbf{w}_k^i - \\ \mu \sum_{j=1}^{p} B_k \mathbf{w}_k^i \left(\mathbf{w}_k^{j^T} B_k \mathbf{w}_k^i - \delta_{ij}\right) \end{array}\right), \quad (7.36)$$

for $i=1,\ldots,p$, the matrix version of which is

$$W_{k+1} = W_k + \eta_k \left(\begin{array}{c} 2A_k W_k - B_k W_k W_k^T A_k W_k - A_k W_k W_k^T B_k W_k - \\ \mu B_k W_k \left(W_k^T B_k W_k - I_p\right) \end{array}\right). \quad (7.37)$$

AL2 Deflation Algorithm

The objective function for the AL2 deflation GEVD algorithm is

$$J\left(\mathbf{w}_k^i; A_k, B_k\right) = -2\mathbf{w}_k^{i^T} A_k \mathbf{w}_k^i + \left(\mathbf{w}_k^{i^T} A_k \mathbf{w}_k^i\right)\left(\mathbf{w}_k^{i^T} B_k \mathbf{w}_k^i\right) +$$

$$2\sum_{j=1}^{i-1} \mathbf{w}_k^{i^T} A_k \mathbf{w}_k^j \mathbf{w}_k^{j^T} B_k \mathbf{w}_k^i + \mu\left(\sum_{j=1}^{i-1}\left(\mathbf{w}_k^{j^T} B_k \mathbf{w}_k^i\right)^2 + \frac{1}{2}\left(\mathbf{w}_k^{i^T} B_k \mathbf{w}_k^i - 1\right)^2\right), \quad (7.38)$$

for $i=1,...,p$. The adaptive GEVD algorithm is

$$W_{k+1} = W_k + \eta_k \left(\begin{array}{c} 2A_k W_k - B_k W_k \text{UT}\left(W_k^T A_k W_k\right) - A_k W_k \text{UT}\left(W_k^T B_k W_k\right) \\ - \mu B_k W_k \text{UT}\left(W_k^T B_k W_k - I_p\right) \end{array} \right). \quad (7.39)$$

AL2 Weighted Algorithm

The objective function for the AL2 weighted generalized eigenvector algorithm is

$$J\left(\mathbf{w}_k^i; A_k, B_k\right) = -2c_i \mathbf{w}_k^{i^T} A_k \mathbf{w}_k^i + c_i \left(\mathbf{w}_k^{i^T} A_k \mathbf{w}_k^i\right)\left(\mathbf{w}_k^{i^T} B_k \mathbf{w}_k^i\right) +$$

$$2\sum_{j=1}^{i-1} c_j \mathbf{w}_k^{i^T} A_k \mathbf{w}_k^j \mathbf{w}_k^{j^T} B_k \mathbf{w}_k^i + \mu \left(\sum_{j=1}^{i-1} c_j \left(\mathbf{w}_k^{i^T} B_k \mathbf{w}_k^i\right)^2 + \frac{c_i}{2}\left(\mathbf{w}_k^{i^T} B_k \mathbf{w}_k^i - 1\right)^2 \right), (7.40)$$

for $i=1,...,p$, where $c_1 > c_2 > ... > c_p > 0$ $(p \leq n)$. The adaptive algorithm is

$$W_{k+1} = W_k + \eta_k \left(\begin{array}{c} 2A_k W_k C - B_k W_k C W_k^T A_k W_k - A_k W_k C W_k^T B_k W_k - \\ \mu B_k W_k C\left(W_k^T B_k W_k - I_p\right) \end{array} \right), \quad (7.41)$$

where $C = diag\left(c_1, ..., c_p\right)$, $p \leq n$.

AL2 Algorithm Python Code

The following Python code works with data X[nDim,nSamples] and Y[nDim,nSamples]:

```
from numpy import linalg as la
A  = np.zeros(shape=(nDim,nDim)) # stores adaptive
                                 correlation matrix
B  = np.zeros(shape=(nDim,nDim)) # stores adaptive
                                 correlation matrix
W2 = 0.1 * np.ones(shape=(nDim,nEA)) # weight vectors of all
                                 algorithms
```

```python
W3 = W2
c = [2.6-0.3*k for k in range(nEA)]
C = np.diag(c)
I = np.identity(nDim)
mu = 1
for epoch in range(nEpochs):
    for iter in range(nSamples):
        cnt = nSamples*epoch + iter
        # Update data correlation matrices A,B with current
        data vectors x,y
        x = X[:,iter]
        x = x.reshape(nDim,1)
        A = A + (1.0/(1 + cnt))*((np.dot(x, x.T)) - A)
        y = Y[:,iter]
        y = y.reshape(nDim,1)
        B = B + (1.0/(1 + cnt))*((np.dot(y, y.T)) - B)
        # Deflated Gradient Descent
        W2 = W2 + (1/(100 + cnt))*(A @ W2 - 0.5 * B @ W2
                        @ np.triu(W2.T @ A \
                        @ W2) - 0.5 * A @ W2 @ np.triu(W2.T
                        @ B @ W2) - \
                        0.5 * mu * B @ W2 @ np.triu((W2.T
                        @ B @ W2) - I))

        # Weighted Gradient Descent
        W3 = W3 + (1/(300 + cnt))*(A @ W3 @ C - 0.5 * B @ W3
                        @ C @ (W3.T @ A \
                        @ W3) - 0.5 * A @ W3 @ C @ (W3.T
                        @ B @ W3) - \
                        0.5 * mu * B @ W3 @ C @ ((W3.T @ B
                        @ W3) - I))
```

7.8 IT GEVD Algorithms

IT Homogeneous Algorithm

The objective function for the information theory homogeneous GEVD algorithm is

$$J\left(\mathbf{w}_k^i; A_k, B_k\right) = \mathbf{w}_k^{i^T} B_k \mathbf{w}_k^i - \ln\left(\mathbf{w}_k^{i^T} A_k \mathbf{w}_k^i\right) + \alpha\left(\mathbf{w}_k^{i^T} B_k \mathbf{w}_k^i - 1\right) +$$
$$2\sum_{j=1, j\neq i}^{p} \beta_j \mathbf{w}_k^{j^T} B_k \mathbf{w}_k^i \tag{7.42}$$

for $i=1,\dots,p$, where $(\alpha, \beta_1, \beta_2, \dots, \beta_p)$ are Lagrange multipliers and $ln(.)$ is logarithm base e. By equating the gradient of (7.42) with respect to $\mathbf{w}_k^i$ to $\mathbf{0}$ and using the constraint $\mathbf{w}_k^{j^T} B_k \mathbf{w}_k^i = \delta_{ij}$, we obtain

$$\alpha = 0 \text{ and } \beta_j = \frac{\mathbf{w}_k^{j^T} A_k \mathbf{w}_k^i}{\mathbf{w}_k^{i^T} A_k \mathbf{w}_k^i}, \tag{7.43}$$

for $j=1,\dots,p$. Replacing $(\alpha, \beta_1, \beta_2, \dots, \beta_p)$ in the gradient of (7.42), we obtain the IT homogeneous adaptive gradient descent algorithm for the generalized eigenvector:

$$\mathbf{w}_{k+1}^i = \mathbf{w}_k^i + \eta_k\left(A_k \mathbf{w}_k^i - \sum_{j=1}^{p} B_k \mathbf{w}_k^j\left(\mathbf{w}_k^{j^T} A_k \mathbf{w}_k^i\right)\right) \Big/ \mathbf{w}_k^{i^T} A_k \mathbf{w}_k^i, \tag{7.44}$$

for $i=1,\dots,p$, whose matrix version is

$$W_{k+1} = W_k + \eta_k\left(A_k W_k - B_k W_k W_k^T A_k W_k\right) \text{DIAG}\left(W_k^T A_k W_k\right)^{-1}, \tag{7.45}$$

where DIAG$[\cdot]$ sets all elements *except* the diagonal of its matrix argument to zero.

IT Deflation Algorithm

The objective function for the IT deflation GEVD algorithm is

$$J\left(\mathbf{w}_k^i; A_k, B_k\right) = {\mathbf{w}_k^i}^T B_k \mathbf{w}_k^i - \ln\left({\mathbf{w}_k^i}^T A_k \mathbf{w}_k^i\right) + \alpha\left({\mathbf{w}_k^i}^T B_k \mathbf{w}_k^i - 1\right) +$$
$$2\sum_{j=1}^{i-1} \beta_j {\mathbf{w}_k^j}^T B_k \mathbf{w}_k^i \tag{7.46}$$

for $i=1,\ldots,p$, where $(\alpha, \beta_1, \beta_2, \ldots, \beta_p)$ are Lagrange multipliers. From (7.43), we obtain the adaptive gradient algorithm:

$$W_{k+1} = W_k + \eta_k\left(A_k W_k - B_k W_k \mathrm{UT}\left(W_k^T A_k W_k\right)\right) \mathrm{DIAG}\left(W_k^T A_k W_k\right)^{-1}. \tag{7.47}$$

IT Weighted Algorithm

The objective function for the IT weighted GEVD algorithm is

$$J\left(\mathbf{w}_k^i\right) = c_i {\mathbf{w}_k^i}^T B_k \mathbf{w}_k^i - c_i \ln\left({\mathbf{w}_k^i}^T A_k \mathbf{w}_k^i\right) + \alpha c_i\left({\mathbf{w}_k^i}^T B_k \mathbf{w}_k^i - 1\right) +$$
$$2\sum_{j=1, j\neq i}^{p} \beta_j c_j {\mathbf{w}_k^j}^T B_k \mathbf{w}_k^i \tag{7.48}$$

for $i=1,\ldots,p$, where $(\alpha, \beta_1, \beta_2, \ldots, \beta_p)$ are Lagrange multipliers. By solving $(\alpha, \beta_1, \beta_2, \ldots, \beta_k)$ and replacing them in the gradient of (7.48), we obtain the adaptive algorithm:

$$W_{k+1} = W_k + \eta_k\left(A_k W_k C - B_k W_k C W_k^T A_k W_k\right) \mathrm{DIAG}\left(W_k^T A_k W_k\right)^{-1}. \tag{7.49}$$

IT Algorithm Python Code

The following Python code works with data X[nDim,nSamples] and
Y[nDim,nSamples]:

```python
from numpy import linalg as la
A  = np.zeros(shape=(nDim,nDim)) # stores adaptive
                                   correlation matrix
B  = np.zeros(shape=(nDim,nDim)) # stores adaptive
                                   correlation matrix
W2 = 0.1 * np.ones(shape=(nDim,nEA)) # weight vectors of all
                                       algorithms
W3 = W2
c = [2.6-0.3*k for k in range(nEA)]
C = np.diag(c)
for epoch in range(nEpochs):
    for iter in range(nSamples):
        cnt = nSamples*epoch + iter
        # Update data correlation matrices A,B with current
        data vectors x,y
        x = X[:,iter]
        x = x.reshape(nDim,1)
        A = A + (1.0/(1 + cnt))*((np.dot(x, x.T)) - A)
        y = Y[:,iter]
        y = y.reshape(nDim,1)
        B = B + (1.0/(1 + cnt))*((np.dot(y, y.T)) - B)
        # Deflated Gradient Descent
        W2 = W2 + (1/(50 + cnt))*(A @ W2 - B @ W2
        @ np.triu(W2.T @ A @ W2)) \
                                  @ inv(np.diag(np.
                                  diagonal(W2.T @ A @ W2)))
```

```
# Weighted Gradient Descent
W3 = W3 + (1/(100 + cnt))*(A @ W3 @ C - B @ W3 @ C
                          @ (W3.T @ A @ W3)) \
                          @ inv(np.diag(np.
                          diagonal(W3.T @ A @ W3)))
```

7.9 RQ GEVD Algorithms

RQ Homogeneous Algorithm

We obtain the objective function for the Rayleigh quotient homogeneous GEVD algorithm from the Rayleigh quotient criterion (7.2) as follows:

$$J\left(\mathbf{w}_k^i; A_k, B_k\right) = -\frac{\mathbf{w}_k^{i^T} A_k \mathbf{w}_k^i}{\mathbf{w}_k^{i^T} B_k \mathbf{w}_k^i} + \alpha\left(\mathbf{w}_k^{i^T} B_k \mathbf{w}_k^i - 1\right) + 2\sum_{j=1, j\neq i}^{p} \beta_j \mathbf{w}_k^{j^T} B_k \mathbf{w}_k^i \quad (7.50)$$

for $i=1,...,p$, where $(\alpha, \beta_1, \beta_2, ..., \beta_p)$ are Lagrange multipliers. By equating the gradient (7.50) with respect to $\mathbf{w}_k^i$ to $\mathbf{0}$, and using the constraint $\mathbf{w}_k^{j^T} B_k \mathbf{w}_k^i = \delta_{ij}$, we obtain

$$\alpha = 0 \text{ and } \beta_j = \frac{\mathbf{w}_k^{j^T} A_k \mathbf{w}_k^i}{\mathbf{w}_k^{i^T} B_k \mathbf{w}_k^i} \text{ for } j=1,...,p. \quad (7.51)$$

Replacing $(\alpha, \beta_1, \beta_2, ..., \beta_p)$ in the gradient of (7.50) and making a small approximation, we obtain the RQ homogeneous adaptive gradient descent algorithm for the generalized eigenvector:

$$\mathbf{w}_{k+1}^i = \mathbf{w}_k^i + \eta_k \left(A_k \mathbf{w}_k^i - \sum_{j=1}^{p} B_k \mathbf{w}_k^j \left(\mathbf{w}_k^{j^T} A_k \mathbf{w}_k^i \right) \right) / \mathbf{w}_k^{i^T} B_k \mathbf{w}_k^i, \quad (7.52)$$

for $i=1,...,p$, whose matrix version is

$$W_{k+1} = W_k + \eta_k \left(A_k W_k - B_k W_k W_k^T A_k W_k \right) \text{DIAG}\left(W_k^T B_k W_k \right)^{-1}. \quad (7.53)$$

RQ Deflation Algorithm

The objective function for the RQ deflation GEVD algorithm is

$$J\left(\mathbf{w}_k^i; A_k, B_k\right) = -\frac{\mathbf{w}_k^{i^T} A_k \mathbf{w}_k^i}{\mathbf{w}_k^{i^T} B_k \mathbf{w}_k^i} + \alpha\left(\mathbf{w}_k^{i^T} B_k \mathbf{w}_k^i - 1\right) + 2\sum_{j=1}^{i} \beta_j \mathbf{w}_k^{j^T} B_k \mathbf{w}_k^i, \quad (7.54)$$

for $i=1,...,p$, where $(\alpha, \beta_1, \beta_2, ..., \beta_p)$ are Lagrange multipliers. By solving $(\alpha, \beta_1, \beta_2, ..., \beta_k)$ and replacing them in the gradient of (7.54), we obtain the adaptive algorithm:

$$W_{k+1} = W_k + \eta_k\left(A_k W_k - B_k W_k \text{UT}\left(W_k^T A_k W_k\right)\right) \text{DIAG}\left(W_k^T B_k W_k\right)^{-1}. \quad (7.55)$$

RQ Weighted Algorithm

The objective function for the RQ weighted GEVD algorithm is

$$J\left(\mathbf{w}_k^i; A_k, B_k\right) = -c_i\frac{\mathbf{w}_k^{i^T} A_k \mathbf{w}_k^i}{\mathbf{w}_k^{i^T} B_k \mathbf{w}_k^i} + \alpha c_i\left(\mathbf{w}_k^{i^T} B_k \mathbf{w}_k^i - 1\right) + 2\sum_{j=1}^{i} \beta_j c_j \mathbf{w}_k^{j^T} B_k \mathbf{w}_k^i, (7.56)$$

for $i=1,...,p$, where $(\alpha, \beta_1, \beta_2, ..., \beta_p)$ are Lagrange multipliers. By solving $(\alpha, \beta_1, \beta_2, ..., \beta_k)$ and replacing them in the gradient of (7.56), we obtain the adaptive algorithm:

$$W_{k+1} = W_k + \eta_k\left(A_k W_k C - B_k W_k C W_k^T A_k W_k\right) \text{DIAG}\left(W_k^T B_k W_k\right)^{-1}. \quad (7.57)$$

RQ Algorithm Python Code

The following Python code works with data X[nDim,nSamples] and Y[nDim,nSamples]:

```
from numpy import linalg as la
A   = np.zeros(shape=(nDim,nDim)) # stores adaptive
                           correlation matrix
```

```python
B  = np.zeros(shape=(nDim,nDim)) # stores adaptive
                              correlation matrix
W2 = 0.1 * np.ones(shape=(nDim,nEA)) # weight vectors of all
                              algorithms
W3 = W2
c = [2.6-0.3*k for k in range(nEA)]
C = np.diag(c)
I  = np.identity(nDim)
for epoch in range(nEpochs):
    for iter in range(nSamples):
        cnt = nSamples*epoch + iter
        # Update data correlation matrices A,B with current
        data vectors x,y
        x = X[:,iter]
        x = x.reshape(nDim,1)
        A = A + (1.0/(1 + cnt))*((np.dot(x, x.T)) - A)
        y = Y[:,iter]
        y = y.reshape(nDim,1)
        B = B + (1.0/(1 + cnt))*((np.dot(y, y.T)) - B)
        # Deflated Gradient Descent
        W2 = W2 + (1/(20 + cnt))*(A @ W2 - B @ W2 @
                            np.triu(W2.T @ A @ W2)) \
                            inv(np.diag(np.diagonal(W2.T
                            @ B @ W2)))

        # Weighted Gradient Descent
        W3 = W3 + (1/(300 + cnt))*(A @ W3 @ C - B @ W3 @ C @
                            (W3.T @ A @ W3)) \
                            @ inv(np.diag(np.
                            diagonal(W3.T @ B @ W3)))
```

7.10 Experimental Results

I generated 1,000 samples (of $\{\mathbf{x}_k\}$ and $\{\mathbf{y}_k\}$) from 10-dimensional Gaussian data (i.e., $n=10$) with the mean zero and covariance given below. The covariance matrix A for $\{\mathbf{x}_k\}$ is obtained from the second covariance matrix in [Okada and Tomita 85] multiplied by 3 as follows:

$$
3\begin{bmatrix}
0.427 & 0.011 & -0.005 & -0.025 & 0.089 & -0.079 & -0.019 & 0.074 & 0.089 & 0.005 \\
0.011 & 5.690 & -0.069 & -0.282 & -0.731 & 0.090 & -0.124 & 0.100 & 0.432 & -0.103 \\
-0.005 & -0.069 & 0.080 & 0.098 & 0.045 & -0.041 & 0.023 & 0.022 & -0.035 & 0.012 \\
-0.025 & -0.282 & 0.098 & 2.800 & -0.107 & 0.150 & -0.193 & 0.095 & -0.226 & 0.046 \\
0.089 & -0.731 & 0.045 & -0.107 & 3.440 & 0.253 & 0.251 & 0.316 & 0.039 & -0.010 \\
-0.079 & 0.090 & -0.041 & 0.150 & 0.253 & 2.270 & -0.180 & 0.295 & -0.039 & -0.113 \\
-0.019 & -0.124 & 0.023 & -0.193 & 0.251 & -0.180 & 0.327 & 0.027 & 0.026 & -0.016 \\
0.074 & 0.100 & 0.022 & 0.095 & 0.316 & 0.295 & 0.027 & 0.727 & -0.096 & -0.017 \\
0.089 & 0.432 & -0.035 & -0.226 & 0.039 & -0.039 & 0.026 & -0.096 & 0.715 & -0.009 \\
0.005 & -0.103 & 0.012 & 0.046 & -0.010 & -0.113 & -0.016 & -0.017 & -0.009 & 0.065
\end{bmatrix}.
$$

The covariance matrix B for $\{\mathbf{y}_k\}$ is obtained from the third covariance matrix in [Okada and Tomita 85] multiplied by 2 as follows:

$$
2\begin{bmatrix}
0.3350 & 0.026 & -0.051 & -0.012 & 0.079 & 0.017 & 0.029 & 0.008 & 0.077 & -0.030 \\
0.026 & 0.091 & 0.011 & -0.010 & 0.006 & -0.014 & -0.002 & -0.023 & 0.011 & 0.035 \\
-0.051 & 0.011 & 0.078 & 0.000 & 0.016 & 0.003 & 0.030 & -0.035 & -0.003 & -0.049 \\
-0.012 & -0.010 & 0.000 & 0.082 & -0.003 & -0.026 & -0.025 & -0.029 & -0.015 & 0.025 \\
0.079 & 0.006 & 0.016 & -0.003 & 0.797 & 0.194 & -0.037 & -0.023 & 0.059 & -0.145 \\
0.017 & -0.014 & 0.003 & -0.026 & 0.194 & 1.500 & 0.014 & -0.104 & 0.114 & -0.229 \\
0.029 & -0.002 & 0.030 & -0.025 & -0.037 & 0.014 & 0.277 & -0.030 & -0.077 & -0.051 \\
0.008 & -0.023 & -0.035 & -0.029 & -0.023 & -0.104 & -0.030 & 0.317 & 0.022 & 0.010 \\
0.077 & 0.011 & -0.003 & -0.015 & 0.059 & 0.114 & -0.077 & 0.022 & 0.538 & 0.034 \\
-0.030 & 0.035 & -0.049 & 0.025 & -0.145 & -0.229 & -0.051 & 0.010 & 0.034 & 0.668
\end{bmatrix}
$$

The generalized eigenvalues of (A,B) are

107.9186, 49.0448, 8.3176, 5.1564, 2.8814, 2.3958, 1.9872, 1.2371, 0.9371, 0.1096.

I computed the first four principal generalized eigenvectors (i.e., eigenvectors corresponding to the largest four eigenvalues) (i.e., $p=4$) by the adaptive algorithms described before. In order to compute the online data sequence $\{A_k\}$, I generated random data vectors $\{\mathbf{x}_k\}$ from the above covariance matrix A. I generated $\{A_k\}$ from $\{\mathbf{x}_k\}$ by using algorithm (2.21) in Chapter 2. Similarly, I generated random data vectors $\{\mathbf{y}_k\}$ from the covariance matrix B and then generated $\{B_k\}$ from $\{\mathbf{y}_k\}$. I computed the correlation matrix $A_{computed}$ and $B_{computed}$ after collecting all 500 samples $\mathbf{x}_k$ and $\mathbf{y}_k$ respectively as

$$A_{computed} = \frac{1}{1000}\sum_{i=1}^{1000}\mathbf{x}_i\mathbf{x}_i^T \text{ and } B_{computed} = \frac{1}{1000}\sum_{i=1}^{1000}\mathbf{y}_i\mathbf{y}_i^T .$$

I referred to the generalized eigenvectors and eigenvalues computed from this A and B by a standard numerical analysis method [Golub and VanLoan 83] as the *actual values*.

I started all algorithms with $\mathbf{w}_0 = 0.1*ONE$, where ONE is a 10 X 4 matrix whose all elements are ones. In order to measure the convergence and accuracy of the algorithms, I computed the direction cosine at k^{th} update of each adaptive algorithm as

$$\text{Direction cosine } (k) = \frac{\left|\mathbf{w}_k^{i\,T}\boldsymbol{\phi}_i\right|}{\|\mathbf{w}_k^i\|\|\boldsymbol{\phi}_i\|}, \tag{7.58}$$

where $\mathbf{w}_k^i$ is the estimated generalized eigenvector of (A_k, B_k) at k^{th} update and $\boldsymbol{\phi}_i$ is the actual i^{th} generalized eigenvector computed from all collected samples by a conventional numerical analysis method.

Figure 7-1 shows the iterates of the OJA algorithms (deflated and weighted) to compute the first two principal generalized eigenvectors of (A_k, B_k). Figure 7-2 shows the same for the XU algorithms, Figure 7-3 for the PF algorithms, Figure 7-4 for the AL1 algorithms, Figure 7-5 for the AL2 algorithms, Figure 7-6 for the IT algorithms, and Figure 7-7 for the RQ algorithms.

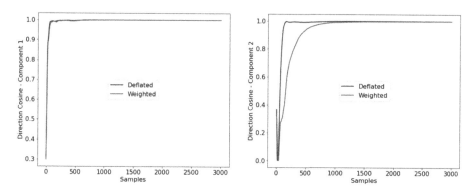

Figure 7-1. *Convergence of the first two principal generalized eigenvectors of (A,B) by the OJA deflation (7.10) and OJA weighted (7.13) adaptive algorithms*

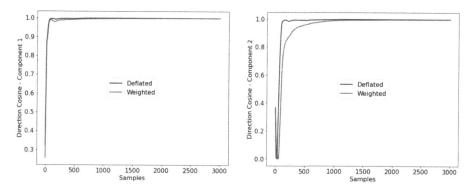

Figure 7-2. *Convergence of the first two principal generalized eigenvectors of (A,B) by the XU deflation (7.18) and XU weighted (7.20) adaptive algorithms*

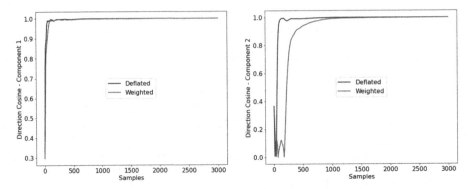

Figure 7-3. *Convergence of the first two principal generalized eigenvectors of (A,B) by the PF deflation (7.24) and PF weighted (7.26) adaptive algorithms*

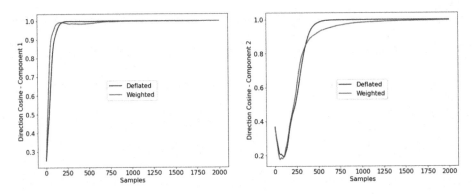

Figure 7-4. *Convergence of the first two principal generalized eigenvectors of (A,B) by the AL1 deflation (7.32) and AL1 weighted (7.34) adaptive algorithms*

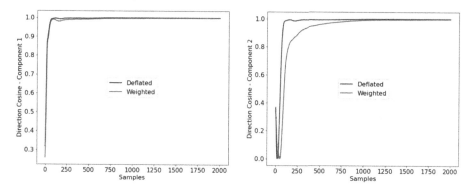

Figure 7-5. *Convergence of the first two principal generalized eigenvectors of (A,B) by the AL2 deflation (7.39) and AL2 weighted (7.41) adaptive algorithms*

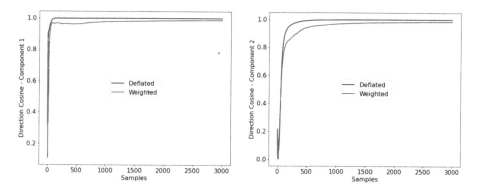

Figure 7-6. *Convergence of the first two principal generalized eigenvectors of (A,B) by the IT deflation (7.47) and IT weighted (7.49) adaptive algorithms*

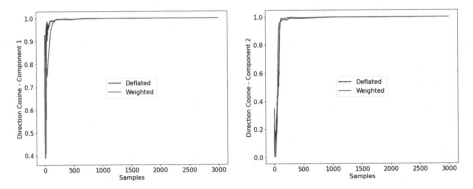

Figure 7-7. *Convergence of the first two principal generalized eigenvectors of (A,B) by the RQ deflation (7.55) and RQ weighted (7.57) adaptive algorithms*

For all algorithms, I used $\eta_k=1/(140+k)$ for the deflation algorithms and $\eta_k=1/(500+k)$ for the weighted algorithms. The diagonal weight matrix C used for the weighted algorithms is DIAG(2.6,2.3,2.0,1.7). I ran all algorithms for *three epochs* of the data, where one epoch means presenting all training data once in random order. I did not show the results for the homogeneous algorithms since the homogeneous method produces a linear combination of the actual generalized eigenvectors of (A,B). Thus, the direction cosines are not indicative of the performance of the algorithms for the homogeneous case.

7.11 Concluding Remarks

Observe that the convergence of all algorithms decreases progressively for the minor generalized eigenvectors and is best for principal generalized eigenvector. This is expected since the convergence of these adaptive algorithms is a function of the relative generalized eigenvalue. Furthermore, the weighted algorithms, which can be implemented in parallel hardware, performed very similarly to the deflation algorithms.

I did the following:

1. For each algorithm, I rated the compute and convergence performance.

2. I skipped the homogeneous algorithms because they are not useful for practical applications since they produce arbitrary rotations of the eigenvectors.

3. Note that $A_k \in \Re^{nXn}$, $B_k \in \Re^{nXn}$ and $W_k \in \Re^{nXp}$. I presented the computation complexity of each algorithm in terms of the matrix dimensions n and p.

4. The convergence performance is determined based on the speed of convergence of the principal and the minor components. I rated convergence in a scale of 1-10 where 10 is the fastest converging algorithm.

5. I skipped the IT and RQ algorithms because they did not perform well compared to the remaining algorithms and the matrix inversion increases computational complexity. See Table 7-3.

Table 7-3. List of Adaptive GEVD Algorithms, Complexity, and Performance

Alg	Type	Adaptive Algorithm $h(W_k, A_k)$	Comments
OJA	Deflation	$A_k W_k - B_k W_k \mathrm{UT}(W_k^T A_k W_k)$	$n^3 p^6$, 6
	Weighted	$A_k W_k C - B_k W_k C W_k^T A_k W_k$	$n^4 p^6$, 6
XU	Deflation	$2A_k W_k - A_k W_k \mathrm{UT}(W_k^T B_k W_k) - B_k W_k \mathrm{UT}(W_k^T A_k W_k)$	$2n^3 p^6$, 8
	Weighted	$2A_k W_k C - B_k W_k C W_k^T A_k W_k - A_k W_k C W_k^T B_k W_k$	$2n^4 p^6$, 8
PF	Deflation	$A_k W_k - \mu B_k W_k \mathrm{UT}(W_k^T B_k W_k - I_p)$	$n^2 p^4$, 7
	Weighted	$A_k W_k C - \mu B_k W_k C(W_k^T B_k W_k - I_p)$	$n^3 p^4$, 7
AL1	Deflation	$A_k W_k - B_k W_k \mathrm{UT}(W_k^T A_k W_k) - \mu B_k W_k \mathrm{UT}(W_k^T B_k W_k - I_p)$	$n^3 p^6 + n^2 p^4$, 9
	Weighted	$A_k W_k C - B_k W_k C W_k^T A_k W_k - \mu B_k W_k C(W_k^T B_k W_k - I_p)$	$n^4 p^6 + n^3 p^4$, 9
AL2	Deflation	$2A_k W_k - B_k W_k \mathrm{UT}(W_k^T A_k W_k) - A_k W_k \mathrm{UT}(W_k^T B_k W_k) - \mu B_k W_k \mathrm{UT}(W_k^T B_k W_k - I_p)$	$2n^3 p^6 + n^2 p^4$, 10
	Weighted	$2A_k W_k C - B_k W_k C W_k^T A_k W_k - A_k W_k C W_k^T B_k W_k - \mu B_k W_k C(W_k^T B_k W_k - I_p)$	$2n^4 p^6 + n^3 p^4$, 10

IT	Deflation	$\left(A_k W_k - B_k W_k \mathrm{UT}\left(W_k^T A_k W_k\right)\right)$	$\mathrm{DIAG}\left(W_k^T A_k W_k\right)^{-1}$	Not applicable
	Weighted	$\left(A_k W_k C - B_k W_k C W_k^T A_k W_k\right)$	$\mathrm{DIAG}\left(W_k^T A_k W_k\right)^{-1}$	Not applicable
RQ	Deflation	$\left(A_k W_k - B_k W_k \mathrm{UT}\left(W_k^T A_k W_k\right)\right)$	$\mathrm{DIAG}\left(W_k^T B_k W_k\right)^{-1}$	Not applicable
	Weighted	$\left(A_k W_k C - B_k W_k C W_k^T A_k W_k\right)$	$\mathrm{DIAG}\left(W_k^T B_k W_k\right)^{-1}$	Not applicable

Observe the following:

1. The OJA algorithm has the least complexity and good performance.

2. The AL2 algorithm has the most complexity and best performance.

3. The AL1 algorithm is the next best after AL2, and PF and XU follow.

The complexity and accuracy tradeoffs will determine the algorithm to use in real-world scenarios. If you can afford the computation, the AL2 algorithm is the best. The XU algorithm is a good balance of complexity and performance.

In summary, I showed 21 algorithms, many of them new, from a common framework with an objective function for each. Note that although I applied the gradient descent technique on these objective functions, I could have applied any other technique of nonlinear optimization such as steepest descent, conjugate direction, Newton-Raphson, or recursive least squares. The availability of the objective functions allows us to derive new algorithms by using new optimization techniques on them and also to perform convergence analyses of the adaptive algorithms.

CHAPTER 8

Real-World Applications of Adaptive Linear Algorithms

In this chapter, I consider real-world examples of linear adaptive algorithms. Some of the best needs for these algorithms arise due to edge computation on devices, which require managing the following:

- Power usage for device-based computation at scale

- Non-stationarity of inputs

- Latency of computation on devices

- Memory and bandwidth of devices

In these cases, there are the following constraints:

- The data arrives as a sequence of random vectors $\{\mathbf{x}_k\}$ or random matrices $\{A_k\}$.

- The data changes with time, causing significant drift of input features whereby the models are no longer effective over time.

© Chanchal Chatterjee 2022
C. Chatterjee, *Adaptive Machine Learning Algorithms with Python*,
https://doi.org/10.1007/978-1-4842-8017-1_8

- The data volume is large and we do not have the device storage, bandwidth, or power to store or upload the data.

- Data dimensionality can be large.

In these circumstances, I will demonstrate how to use the linear adaptive algorithms to manage the device's power, memory, and bandwidth in order to maintain accuracy of the pretrained models. The examples I will cover are the following:

- Calculating feature drift of incoming data and detecting training-serving skew [Kaz Sato et al. 21] ahead of time

- Adapting to incoming data drift and calculating features that best fit the data

- Compressing incoming data into features for use in new model creation

- Calculating anomalies in incoming data so that good clean data is used by the models

In these examples, I used data from the following repository: Publicly Real-World Datasets to Evaluate Stream Learning Algorithms. This dataset represents real-world streaming non-stationarity data [Vinicius Souza et al. 20].

Note that besides the examples discussed in this chapter, I have considered many other practical examples of adaptive examples throughout the book, such as

- Handwritten character recognition with adaptive mean

- Anomaly detection with adaptive median

- Data representation feature computation

- Data classification feature computation

8.1 Detecting Feature Drift

As the underlying statistical properties of the incoming data changes with time, the models used for machine learning decay in performance. Early detection of feature drift and retraining the models maintains the accuracy of the machine learning solution.

INSECTS-incremental_balanced_norm Dataset: Eigenvector Test

The dataset name is `INSECTS-incremental_balanced_norm.arff`. This dataset has 33 components. It has gradually increasing components causing the feature drift shown in Figure 8-1.

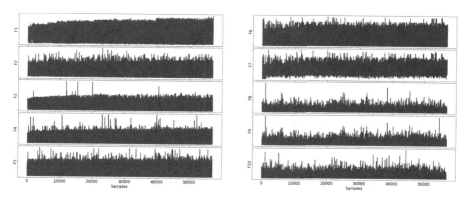

Figure 8-1. *Non-stationary multi-dimensional real-world data with incremental drift*

Adaptive EVD of Semi-Stationary Components

I dropped the drift components and used the EVD linear adaptive algorithm (5.13) from Chapter 5, shown below, on the remaining components:

$$W_{k+1} = W_k + \eta_k \left(2A_k W_k - A_k W_k \mathrm{UT}\left(W_k^T W_k\right) - W_k \mathrm{UT}\left(W_k^T A_k W_k\right)\right). \quad (5.13)$$

219

Note that the remaining components are a lot more stable, but some non-stationarity still exists. For each input sample of the sequence, I plotted the norms of the first four principal eigenvectors to demonstrate the quality of convergence of these eigenvectors; see Figure 8-2.

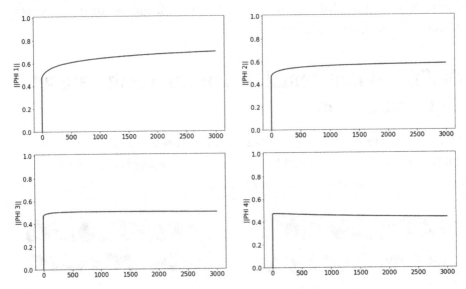

Figure 8-2. *Norms of the first four eigenvectors for the adaptive EVD algorithm (5.13) on stationary data*

The first four eigenvector norms converge rapidly to stable values with streaming samples. The upward horizontal slopes of the curves indicate stable convergence. The slight downward slopes of the third and fourth principal eigenvectors show a slight non-stationarity in the data. But the data is largely stable and stationary. We can conclude that the features are consistent with the current machine learning model and no model changes are necessary.

The following Python code works on multidimensional data dataset2 [nDim, nSamples]:

```python
# Adaptive algorithm
from numpy import linalg as la
nSamples = dataset2.shape[0]
nDim = dataset2.shape[1]
A = np.zeros(shape=(nDim,nDim)) # stores adaptive
correlation matrix
N = np.zeros(shape=(1,nDim)) # stores eigen norms
W = 0.1 * np.ones(shape=(nDim,nDim)) # stores adaptive
eigenvectors
for iter in range(nSamples):
    cnt = iter + 1
    # Update data correlation matrix A with current data
    vector x
    x = np.array(dataset2.iloc[iter])
    x = x.reshape(nDim,1)
    A = A + (1.0/cnt)*((np.dot(x, x.T)) - A)
    etat = 1.0/(25 + cnt)
    # Deflated Gradient Descent
    W = W + etat*(A @ W - 0.5*W @ np.triu(W.T @ A @ W) - \
                  0.5*A @ W @ np.triu(W.T @ W))
    newnorm = la.norm(W, axis=0)
    N = np.vstack([N, newnorm])
```

Adaptive EVD of Non-Stationary Components

I next used the non-stationary components of the data to detect feature drift with the adaptive algorithms. I used the same adaptive EVD algorithm (5.13) and plotted the norms of the first four principal eigenvectors for each data sample, as shown in Figure 8-3.

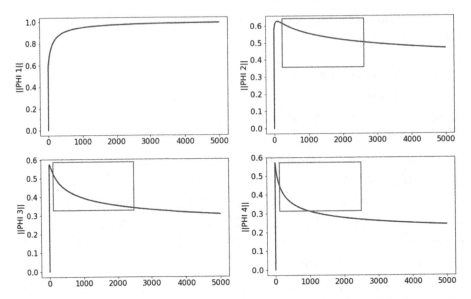

Figure 8-3. *Norms of the first four eigenvectors for the adaptive EVD algorithm (5.13) on non-stationary data*

You can clearly see that the second through fourth eigenvectors diverge, indicated by the downward slopes of the graphs, showing the feature drift early in the sequence. The downward slope of the second through fourth eigenvectors indicates the gradual drift of the features. This result shows that the features are drifting from the original ones used to build the machine learning model.

I used the same Python code I used on the stationary data in Section 8.1.1.

INSECTS-incremental-abrupt_balanced _norm Dataset

The dataset name is INSECTS-incremental_abrupt_balanced_norm. arff. This dataset has repeated abrupt changes in features, as shown in Figure 8-4.

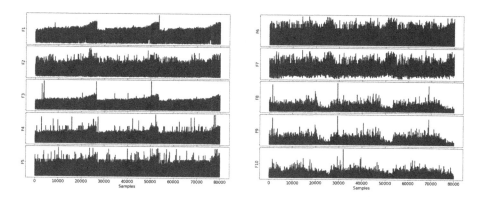

Figure 8-4. *Non-stationary multi-dimensional real-world data with periodic abrupt drift*

I used the same adaptive EVD algorithm (5.13) and observed the norms of the first four principal eigenvectors and plotted them for each data sample, as shown in Figure 8-5.

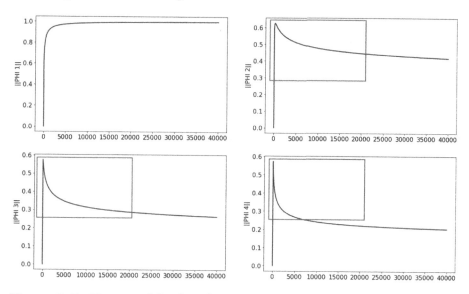

Figure 8-5. *Norms of the first four eigenvectors for the adaptive EVD algorithm (5.13) on non-stationary data*

Once again, the second through fourth eigenvectors diverge, indicating feature drift early in the data sequence. The downward slope of the second through fourth eigenvectors detects drift of the features early in the sequence.

I used the same Python code I used on the stationary data in Section 8.1.1.

Electricity Dataset

The dataset name is `elec.arff`. This dataset has a variety of non-stationary components, as shown in Figure 8-6.

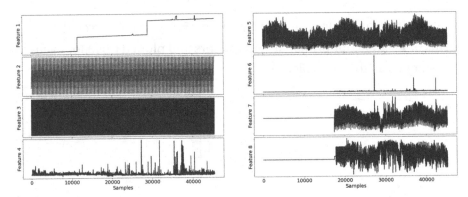

Figure 8-6. *Non-stationary multi-dimensional real-world data with abrupt drifts and trends*

The adaptive EVD algorithm (5.13) gave us the first two principal eigenvectors shown in Figure 8-7, indicating non-stationarity early in the sequence.

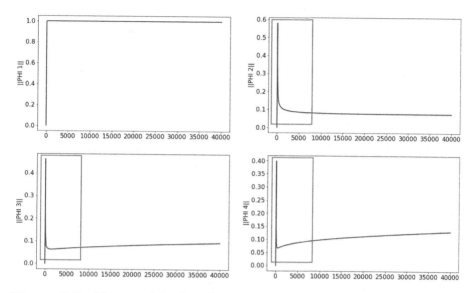

Figure 8-7. *Norms of the first four eigenvectors for the adaptive EVD algorithm (5.13) on non-stationary data*

Figure 8-7 shows the rapid drop in norms of the second through fourth eigenvectors computed by the adaptive algorithm (5.13). This example shows that large non-stationarity in the data is signaled very quickly by massive drops in norms right at the start of the data sequence.

I used the same Python code I used on the stationary data in Section 8.1.1.

8.2 Adapting to Incoming Data Drift

While it is important to detect drift of non-stationary data, it is also important for our algorithms to adapt to data drift quickly. See the example of *simulated data* in Figure 8-8 that abruptly changes to a different underlying statistic after 500 samples.

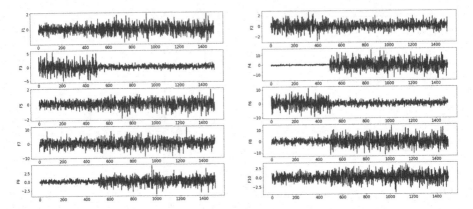

Figure 8-8. *Non-stationary multi-dimensional simulated data with an abrupt change after 500 samples*

I used the adaptive steepest descent algorithm (6.8) to compute the principal eigenvectors. The adaptive algorithm helps us adapt to this abrupt change and recalculate the underlying PCA statistics—in this case, the first two principal eigenvectors of the data. See Figure 8-9.

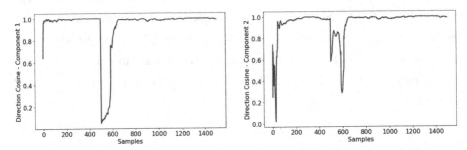

Figure 8-9. *The adaptive steepest descent algorithm (6.8) rapidly adapts to abrupt non-stationary data for principal eigenvector computation.*

The Python code for this algorithm is given in Section 6.3.

8.3 Compressing High Volume and High Dimensional Data

When the incoming data volume is large, it can be prohibitively difficult to store such data on the device for training machine learning models. The problem is further complicated if the data is high dimensional, like 100+ dimensions. In such circumstances, we want to compress the data into batches and store the sequence of feature vectors for future use for machine learning training.

In this example, I used the open source `gassensor.arff` data. The dataset has 129 components/dimensions and 13,910 samples. I used the adaptive EVD algorithm (5.13) to compute the first 16 principal components $[\phi_1 \, \phi_2 \dots \phi_{16}]$. I reconstructed the data back from these 16-dimensional principal components. In Figure 8-10, the left column is the original data and the right column is the reconstructed data. Clearly, they look quite similar and there is an 8x data compression.

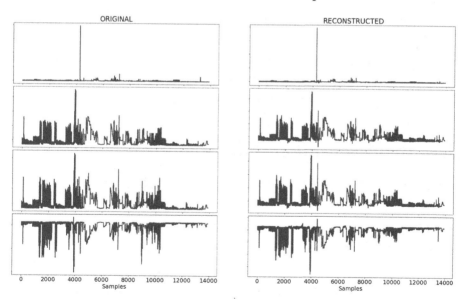

Figure 8-10. *Original (left) and reconstructed (right) data with 8x compression using the adaptive EVD algorithm (5.13)*

227

The Python code is given in Section 5.4 and below for `dataset[nDim,nSamples]`:

```python
from numpy import linalg as la
nSamples = dataset.shape[0]
nDim = dataset.shape[1]
nEA = 16
A = np.zeros(shape=(nDim,nDim)) # stores adaptive
correlation matrix
N = np.zeros(shape=(1,nEA)) # stores eigen norms
W = 0.1 * np.ones(shape=(nDim,nEA)) # stores adaptive eigenvectors
for iter in range(nSamples):
    cnt = iter + 1
    # Update data correlation matrix A with current sample x
    x = np.array(dataset.iloc[iter])
    x = x.reshape(nDim,1)
    A = A + (1.0/cnt)*((np.dot(x, x.T)) - A)
    etat = 1.0/(500 + cnt)
    # Deflated Gradient Descent
    W = W + etat*(A @ W - 0.5*W @ np.triu(W.T @ A @ W) - \
                0.5*A @ W @ np.triu(W.T @ W))
```

Data Representation (PCA) Features

Let's demonstrate the success of the adaptive algorithms by comparing the eigenvalues and eigenvectors of the batch correlation matrix computed by conventional method with the adaptive algorithm (5.13) at each adaptive step for the first four principal eigenvectors.

In order to measure the convergence and accuracy of the adaptive algorithm, I computed the direction cosine at k^{th} update eigenvector of each adaptive algorithm as

$$\text{Direction cosine } (k) = \left|{\mathbf{w}_k^i}^T \phi_i\right| \Big/ \|\phi_i\| \, \|\mathbf{w}_k^i\|,$$

where $\mathbf{w}_k^i$ is the estimated eigenvector of A_k at k^{th} update and $\boldsymbol{\phi}_i$ is the actual i^{th} principal eigenvector computed from all collected samples by a conventional numerical analysis method. I measured the error of the eigenvalues at the kth update as follows:

$$\text{Abs error}\,(k) = \left| d_k^i - \lambda_k^i \right|,$$

where d_k^i is the estimated eigenvalue of A_k at k^{th} update and λ_k^i is the actual i^{th} principal eigenvalue computed from all collected samples by a conventional numerical analysis method.

I did the following:

1. I plotted the direction cosines of the eigenvectors with the batch eigenvectors for each adaptive step. See Figure 8-11.

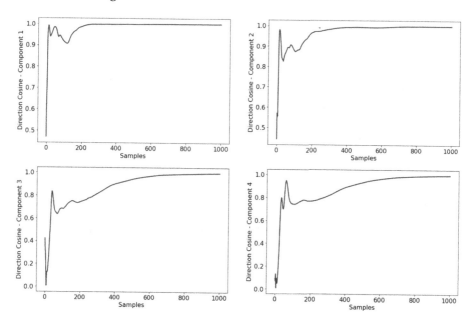

Figure 8-11. Direction cosines of the first four principal eigenvectors with the adaptive algorithm (5.13) (ideal value = 1)

2. I plotted the errors of the eigenvalues with the batch
    eigenvalues for each adaptive step. See Figure 8-12.

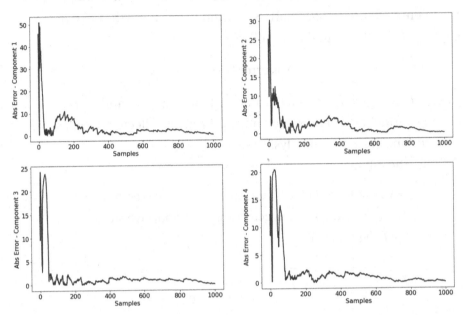

Figure 8-12. *Absolute error of the first four principal eigenvalues with the adaptive algorithm (5.13)*

The final results are

- Actual eigenvalues: [57.865, 36.259, 27.087, 21.833]

- Adaptive eigenvalues: [57.857, 36.251, 27.054, 21.881]

These results demonstrate that the adaptive algorithm accurately computed the PCA features, which will create an 8x data compression in place of the voluminous raw data.

The Python code used here is same as in the section before this one.

8.4 Detecting Feature Anomalies

Here I used adaptive linear algorithms to detect anomalies in data.

Yahoo Real Dataset

The dataset is `real_data.csv` and is derived from Yahoo Research Webscope: S5 - A Labeled Anomaly Detection Dataset [Yahoo Research Webscope]. Detecting anomalies in machines is an important machine learning task for the manufacturing market. I collected 8-dimensional data from machines, which have occasional anomalous readings. One method of detecting anomalies is to calculate the deviation of the current sample from its running median and compare it against a statistic.

I used the adaptive median algorithm (2.20) to calculate the running median and a simple statistic to detect anomalies. See the results in Figure 8-13. I plotted the data in blue, the adaptive median in green, and the anomalies in red.

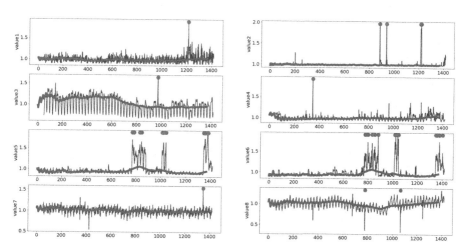

Figure 8-13. *Yahoo Webscope multidimensional real dataset with anomalies in several components detected by the adaptive median algorithm (2.20)*

The following Python code is used on multidimensional dataset
X[nDim, nSamples]:

```python
from numpy import linalg as la
nSamples = X.shape[0]
nDim = X.shape[1]
w = np.zeros(shape=(nDim,1)) # stores adaptive eigenvectors
anam = np.zeros(shape=(nDim,nSamples))
mdks = np.zeros(shape=(nDim,nSamples))
for iter in range(nSamples):
    cnt = iter + 1
    # current data vector x
    x = np.array(X.iloc[iter])
    x = x.reshape(nDim,1)
    #Eq.2.20
    w = w + (1/(1 + iter)) * np.sign(x - w)
    mdks[:,iter] = w.T
    y = (np.abs(x-w) > 0.5*w) # Anomaly detection threshold
    anam[:,iter] = y.T
```

NOAA Dataset

The dataset name is NOAA.arff. The data sequence has eight components. Component F3 has an anomalous spike, as shown in Figure 8-14.

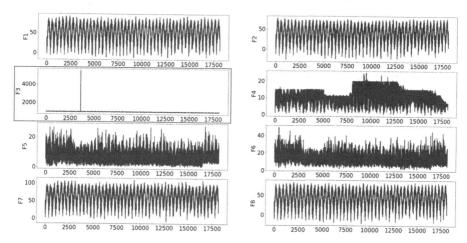

Figure 8-14. *NOAA real-world dataset with an anomaly in one component only, detected by the adaptive median algorithm (2.20)*

I used the adaptive median detection algorithm (2.20) on all components of this dataset. I used a simple algorithm of testing to determine an anomaly:

Anomaly = ABS (Sample_Value – Adaptive_Median) > 4 x Adaptive_Median.

Figure 8-15 shows the data in blue, the adaptive median in green, and the anomalies in red. Clearly the adaptive median algorithm (2.20) detects the anomaly accurately.

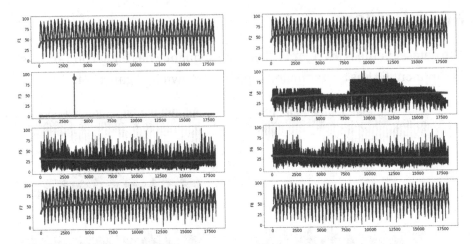

Figure 8-15. *Anomaly in the NOAA dataset detected with the adaptive median algorithm (2.20)*

The Python code used here is same as the previous example except for the detection threshold:

```
y = (np.abs(x-w) > 4*w) # Anomaly detection threshold.
```

References

[1]. E. Oja, "A Simplified Neuron Model as a Principal Component Analyzer", *Journ. of Mathematical Biology*, Vol. 15, pp. 267-273, 1982.

[2]. E. Oja, J. Karhunen, "An Analysis of Convergence for a Learning Version of the Subspace Method", *Journ. Of Math Anal. And Appl.*, 91, 102-111, 1983.

[3]. E. Oja and J. Karhunen, "An Analysis of Convergence for a Learning Version of the Subspace Method", *Journ. of Mathematical Analysis and Applications*, Vol. 91, pp. 102-111, 1983.

[4]. E. Oja and J. Karhunen, "On Stochastic Approximation of the Eigenvectors and Eigenvalues of the Expectation of a Random Matrix", *Journ. of Math. Anal. Appl.*, Vol. 106, pp. 69-84, 1985.

[5]. D. W. Tank and J. J. Hopfield, "Simple neural optimization networks: an A/D converter, signal decision circuit, and a linear programming circuit", *IEEE Trans. Circuits Syst.*, CAS-33, pp. 533-541, 1986.

[6]. H. Bourland and Y. Kamp, "Auto-association by multilayer perceptrons and singular value decomposition", *Biological Cybernetics*, Vol. 59, pp. 291-294, 1988.

© Chanchal Chatterjee 2022
C. Chatterjee, *Adaptive Machine Learning Algorithms with Python*,
https://doi.org/10.1007/978-1-4842-8017-1

REFERENCES

[7]. J. F. Yang and M. Kaveh, "Adaptive eigensubspace algorithm for direction or frequency estimation and tracking", *IEEE Trans. Acoust., Speech, Signal Processing*, vol. 36, no. 2, pp. 241-251, 1988.

[8]. H. Asoh and N. Otsu, "Nonlinear data analysis and multilayer perceptrons", *IEEE INNS Int'l Joint Conf. on Neural Networks*, Vol. 2, pp. 411-415, 1989.

[9]. P. Baldi and K. Hornik, "Neural Networks and Principal Component Analysis: Learning from Examples Without Local Minima", *Neural Networks*, Vol. 2, pp. 53-58, 1989.

[10]. Y. Chauvin, "Principal Component Analysis by Gradient Descent on a Constrained Linear Hebbian Cell", *Proc. Joint Int. Conf. On Neural Networks*, San Diego, CA, Vol. I, pp. 373-380, 1989.

[11]. X. Yang, T. K. Sarkar, and E. Arvas, "A Survey of Conjugate Gradient Algorithms for Solution of Extreme Eigen-Problems of a Symmetric Matrix", *IEEE Transactions on Acoustics, Speech and Signal Processing*, Vol. 37, No. 10, pp. 1550-1556, 1989.

[12]. E. Oja, "Neural networks, principal components, and subspaces", *International Journal Of Neural Systems*, Vol. 1, No. 1, pp. 61-68, 1989.

[13]. J. Rubner and P. Tavan, "A Self-Organizing Network for Principal Component Analysis", *Europhysics Letters*, Vol. 10, No. 7, pp. 693-698, 1989.

[14]. T. K. Sarkar and X. Yang, "Application of the Conjugate Gradient and Steepest Descent for Computing the Eigenvalues of an Operator", *Signal Processing*, Vol. 17, pp. 31-38, 1989.

[15]. T. D. Sanger, "Optimal Unsupervised Learning in a Single-Layer Linear Feedforward Neural Network", *Neural Networks*, Vol. 2, pp. 459-473, 1989.

[16]. P. Foldiak, "Adaptive Network for Optimal Linear Feature Extraction", *Proc. IJCNN*, Washington, pp. 401-405, 1989.

[17]. J. Rubner and K. Schulten, "Development of Feature Detectors by Self-Organization - A Network Model", *Biological Cybernetics*, Vol. 62, pp. 193-199, 1990.

[18]. S. Y. Kung and K. I. Diamantaras, "A neural network learning algorithm for adaptive principal component extraction (APEX)", *Int'l Conf. on Acoustics, Speech and Signal Proc.*, Albuquerque, NM, pp. 861-864, 1990.

[19]. R. W. Brockett, "Dynamical systems that sort lists, diagonalize matrices, and solve linear programming problems", *Linear Algebra and Its Applications*, 146, pp. 79-91, 1991.

[20]. L. Xu, "Least MSE Reconstruction for Self-Organization: (II) Further Theoretical and Experimental Studies on One Layer Nets", *Proc. Int'l Joint Conf. on Neural Networks*, Singapore, pp. 2368-2373, 1991.

[21]. W. Ferzali and J. G. Proakis, "Adaptive SVD Algorithm With Application to Narrowband Signal Tracking", *SVD and Signal Processing, II*: Algorithms, Analysis and Applications, R. J. Vaccaro (Editor) Elsevier Science Publishers B.V., pp. 149-159, 1991.

REFERENCES

[22]. J. Karhunen and J. Joutsensalo, "Frequency estimation by a Hebbian subspace learning algorithm", *Artificial Neural Networks*, T. Kohonen, K. Makisara, O. Simula and J. Kangas (Editors), Elsiver Science Publishers, North-Holland, pp. 1637-1640, 1991.

[23]. J. A. Sirat, "A fast neural algorithm for principal component analysis and singular value decomposition", *Int'l Journ. of Neural Systems*, Vol. 2, Nos. 1 and 2, pp. 147-155, 1991.

[24]. W. R. Softky and D. M. Kammen, "Correlations in High Dimensional or Asymmetric Data Sets: Hebbian Neuronal Processing", *Neural Networks*, Vol. 4, pp. 337-347, 1991.

[25]. T. Leen, "Dynamics of learning in linear feature-discovery networks", *Network*, Vol. 2, pp. 85-105, 1991.

[26]. C. M. Kuan and K. Hornik, **"Convergence of learning algorithms with constant learning rates"**, *IEEE Transactions on Neural Networks*, pp. 484-489, Vol. 2, No. 5, 1991.

[27]. H. Kuhnel and P. Tavan, "A network for discriminant analysis", *Artificial Neural Networks*, T. Kohonen, K. Makisara, O. Simula, J. Kangas (Editors), Amsterdam, Netherlands: Elsevier, 1991.

[28]. A. Cichocki and R. Unbehauen, "Neural networks for computing eigenvalues and eigenvectors", *Biol. Cybern.*, Vol. 68, pp. 155-164, 1992.

[29]. E. Oja, "Principal Components, Minor Components, and Linear Neural Networks", *Neural Networks*, Vol. 5, pp. 927-935, 1992.

[30]. E. Oja, H. Ogawa, and J. Wangviwattana, "Principal Component Analysis by Homogeneous Neural Networks, Part I: The Weighted Subspace Criterion", *IEICE Trans. Inf. & Syst.*, Vol. E75-D, No. 3, pp. 366-375, 1992.

[31]. E. Oja, H. Ogawa, and J. Wangviwattana, "Principal Component Analysis by Homogeneous Neural Networks, Part II: Analysis and Extensions of the Learning Algorithms", *IEICE Trans. Inf. & Syst.*, Vol. E75-D, No. 3, pp. 376-381, 1992.

[32]. M. Moonen, P. VanDooren, and J. Vandewalle, "A Singular Value Decomposition Updating Algorithm For Subspace Tracking", *Siam Journ. Matrix Anal. Appl.*, Vol. 13, No. 4, pp. 1015-1038, October 1992.

[33]. G. W. Stewart, "An updating algorithm for subspace tracking", *IEEE Trans. Signal Proc.*, 40, pp. 1535-1541, 1992.

[34]. K. Gao, M. O. Ahmad, and M. N. S. Swamy, **"Learning algorithm for total least-squares adaptive signal processing",** *Electronics, Letters*, Vol. 28, No. 4, pp. 430 – 432, 1992.

[35]. J. Mao and A. K. Jain, "Discriminant Analysis Neural Networks", *IEEE Int'l Conf. on Neural Networks*, Vol.1, pp. 300-305, San Francisco, CA, March 1993.

[36]. M. Plumbley, "Efficient Information Transfer and anti-Hebbian Neural Networks", Neural Networks, Vol. 6, pp. 823-833, 1993.

[37]. L. Xu, "Least Mean Square Error Reconstruction Principle for Self Organizing Neural Nets", Neural Networks, Vol. 6, pp. 627-648, 1993.

[38]. Z. Fu and E. M. Dowling, "Conjugate Gradient Projection Subspace Tracking", Proc. 1994 Conf. On Signals, Systems and Computers, Pacific Grove, CA, Vol. 1, pp.612-618, 1994.

[39]. J. Karhunen, "Stability of Oja's PCA Subspace Rule", Neural Computation, Vol. 6, pp. 739-747, 1994.

[40]. G. Mathew and V. U. Reddy, "Development and analysis of a neural network approach to Pisarenko's harmonic retrieval method", IEEE Trans. Signal Processing, Vol.42, No.3, pp. 663-667, 1994.

[41]. W-Y. Yan, U. Helmke, and J. B. Moore, "Global Analysis of Oja's Flow for Neural Networks", IEEE Transactions on Neural Networks, Vol. 5, No. 5, 1994.

[42]. K. I. Diamantaras, "Multilayer Neural Networks for Reduced-Rank Approximation", IEEE Transactions on Neural Networks, Vol. 5, No. 5, 1994.

[43]. K. Matsuoka and M. Kawamoto, "A Neural Network that Self-Organizes to Perform Three Operations Related to Principal Component Analysis", *Neural Networks*, Vol. 7, No. 5, pp. 753-765, 1994.

[44]. S. Y. Kung and K. I. Diamantaras, J.S.Taur, "Adaptive Principal Component Extraction (APEX) and Applications", *IEEE Trans. On Signal Proc.*, 42, pp. 1202-1217, 1994.

[45]. G. Mathew and V. U. Reddy, "Orthogonal Eigensubspace Estimation Using Neural Networks", *IEEE Trans. Signal Processing*, Vol.42, No.7, pp. 1803-1811, 1994.

[46]. K. Gao, M. O. Ahmad, and M. N. S. Swamy, "A Constrained Anti-Hebbian Learning Algorithm for Total Least-Squares Estimation with Applications to Adaptive FIR and IIR Filtering", *IEEE Trans. Circuits and Systems II*, Vol. 41, No. 11, pp. 718-729, 1994.

[47]. H. Chen and R. Liu, "An On-Line Unsupervised Learning Machine for Adaptive Feature Extraction", *IEEE Trans. Circuits and Systems II*, Vol. 41, pp. 87-98, 1994.

[48]. P. F. Baldi and K. Hornik, "Learning in Linear Neural Networks: A Survey", *IEEE Transactions on Neural Networks*, Vol. 6, No. 4, pp. 837-858, 1995.

[49]. S. Bannour and M. R. Azimi-Sadjadi, "Principal Component Extraction Using Recursive Least Squares Learning", *IEEE Transactions on Neural Networks*, Vol. 6, No. 2, pp. 457-469, 1995.

[50]. S. Choi, T. K. Sarkar, and J. Choi, "Adaptive antenna array for direction-of -arrival estimation utilizing the conjugate gradient method", *Signal Processing*, vol. 45, pp.313-327, 1995.

[51]. Q.Zhang and Y-W. Leung, "Energy Function for the One-Unit Oja Algorithm", *IEEE Transactions on Neural Networks*, Vol. 6, No. 5, pp. 1291-1293, 1995.

[52]. W-Y. Yan, U. Helmke, and J. B. Moore, "Global Analysis of Oja's Flow for Neural Networks", *IEEE Trans. on Neural Networks*, Vol. 5, No. 5, pp. 674-683, 1994.

[53]. B. Yang, "Projection Approximation Subspace Tracking", *IEEE Transactions on Signal Processing*, Vol. 43, No. 1, pp. 95-107, 1995.

[54]. J. F. Yang and C. L. Lu, "Combined Techniques of Singular Value Decomposition and Vector Quantization for Image Coding", *IEEE Transactions On Image Processing*, Vol. 4, No. 8, pp. 1141-1146, 1995.

[55]. J. L. Wyatt, Jr. and I. M. Elfadel, "Time-Domain Solutions of Oja's Equations", *Neural Computation*, Vol. 7, pp. 915-922, 1995.

[56]. L. Xu and A. L. Yuille, "Robust Principal Component Analysis by Self-Organizing Rules Based on Statistical Physics Approach", *IEEE Transactions on Neural Networks*, Vol. 6, No. 1, pp. 131-143, 1995.

[57]. Z. Fu and E. M. Dowling, "Conjugate Gradient Eigenstructure Tracking for Adaptive Spectral Estimation", *IEEE Transactions on Signal Processing*, Vol. 43, No. 5, pp. 1151-1160, 1995.

[58]. M. D. Plumbley, "Lyapunov Functions for Convergence of Principal Component Algorithms", *Neural Networks*, Vol. 8, No. 1, pp. 11-23, 1995.

[59]. J. Karhunen and J. Joutsensalo, "Generalizations of Principal Component Analysis, Optimization Problems, and Neural Networks", *Neural Networks*, Vol. 8, No. 4, pp. 549-562, 1995.

[60]. G. Mathew, V. U. Reddy, and S. Dasgupta, "Adaptive Estimation of Eigensubspace", *IEEE Transactions on Signal Processing*, Vol. 43, No. 2, pp. 401-411, 1995.

[61]. L-H. Chen and S. Chang, "An Adaptive Learning Algorithm for Principal Component Analysis", *IEEE Transactions on Neural Networks*, Vol. 6, No. 5, 1995.

[62]. P. Strobach, "Fast Recursive Eigensubspace Adaptive Filters", *Proc. ICASSP-95*, Detroit, MI, pp. 1416-1419, 1995.

[63]. C. Chatterjee and V. P. Roychowdhury, "Self-Organizing and Adaptive Algorithms for Generalized Eigen-Decomposition", *Proceedings Advances in Neural Information Processing Systems (NIPS) Conference '96*, Denver, Colorado, November 1996.

[64]. C. Chatterjee and V. P. Roychowdhury, "Self-Organizing Neural Networks for Class-Separability Features", *Proceedings IEEE International Conference on Neural Networks (ICNN '96)*, Washington D.C., June 3-6, pp. 1445-1450, Vol 3, 1996.

[65]. C. Chatterjee, "Adaptive Self-Organizing Neural Networks for Matrix Eigen-Decomposition Problems and their Applications to Feature Extraction", *Ph.D. Dissertation, Purdue University, School of Electrical Engineering*, West Lafayette, IN, May 1996.

REFERENCES

[66]. G. Mathew and V. U. Reddy, "A quasi-Newton adaptive algorithm for generalized symmetric eigenvalue problem", *IEEE Trans. Signal Processing*, vol. 44, no.10, pp. 2413-2422, 1996.

[67]. W. Kasprzak and A. Cichocki, "Recurrent Least Squares Learning for Quasi-Parallel Principal Component Analysis", *ESANN*, Proc. D'facto Publ., pp. 223-228, 1996.

[68]. W. Skarbek, A. Cichocki, and W. Kasprzak, "Principal Subspace Analysis for Incomplete Image Data in One Learning Epoch", *NNWorld*, Vol. 6, No. 3, Prague, pp. 375-382, 1996.

[69]. K. I. Diamantaras, S. Y. Kung, *Principal Component Neural Networks: Theory and Applications*, John Wiley & Sons, 1996.

[70]. C. Chatterjee, V. P. Roychowdhury, M. D. Zoltowski, and J. Ramos, "Self-Organizing and Adaptive Algorithms for Generalized Eigen-Decomposition", *IEEE Transactions on Neural Networks*, Vol. 8, No. 6, pp. 1518-1530, November 1997.

[71]. C. Chatterjee and V. P. Roychowdhury, "On Self-Organizing Algorithms and Networks for Class-Separability Features", *IEEE Transactions on Neural Networks*, Vol. 8, No. 3, pp. 663-678, May 1997.

[72]. C. Chatterjee and V. P. Roychowdhury, "An Adaptive Stochastic Approximation Algorithm for Simultaneous Diagonalization of Matrix Sequences with Applications", *IEEE Transactions on Pattern Analysis and Machine Intelligence*, Vol. 19, No. 3, pp. 282-287, March 1997.

[73]. C. Chatterjee and V. P. Roychowdhury, "Adaptive Algorithms for Eigen-Decomposition and Their Applications in CDMA Communication Systems", *Proceedings 31ᵗʰ Asilomar Conf. on Signals, Systems and Computers*, Nov. 2-5, Pacific Grove, CA, pp. 1575-1580, Vol 2, 1997.

[74]. C. Chatterjee and V. P. Roychowdhury, "Convergence Study of Principal Component Analysis Algorithms", *Proceedings IEEE International Conference on Neural Networks (ICNN '97)*, 1997, Houston, Texas, June 9-12, pp. 1798-1803, Vol. 3, 1997.

[75]. T. Chen, "Modified Oja's algorithms for principal subspace and minor subspace extraction", *Neural Processing Letters*, 5, pp. 105-110, 1997.

[76]. K. I. Diamantaras and M. G. Strintzis, "Noisy PCA theory and application in filter bank codec design", *Proc. IEEE International Conference on Acoustics, Speech, and Signal Processing*, Los Alamitos, CA, USA. pp. 3857-3860, 1997.

[77]. W. Zhu and Y. Wang, "Regularized Total Least Squares Reconstruction for Optical Tomographic Imaging Using Conjugate Gradient Method", *Proc. Int'l Conf. On Image Processing*, Santa Barbara, CA, Vol. 1, pp. 192-195, 1997.

[78]. F. L. Luo and R. Unbehauen, "A minor subspace analysis algorithm", *Neural Networks*, Vol. 8, No. 5, pp. 1149-1155, 1997.

[79]. E. Luo, R. Unbehauen, A. Cichocki, "A minor component analysis algorithm", *Neural Networks*, Vol. 10, No. 2, pp. 291-297, 1997.

[80]. P. Strobach, "Bi-Iteration SVD Subspace Tracking Algorithms", *IEEE Trans. on Signal Processing*, Vol. 45, No. 5, pp. 1222-1240, 1997.

[81]. C. Chatterjee and V. P. Roychowdhury, "On Hetero-Associative Neural Networks and Adaptive Interference Cancellation", *IEEE Transactions on Signal Processing*, Vol. 46, No. 6, pp. 1769-1776, June 1998.

[82]. C. Chatterjee, V. P. Roychowdhury, and E. K. P. Chong, "On Relative Convergence Properties of Principal Component Analysis Algorithms", *IEEE Transactions on Neural Networks*, Vol. 9, No. 2, pp. 319-329, March 1998.

[83]. T. Chen, Y. Hua, and W-Y. Yan, "Global Convergence of Oja's Subspace Algorithm for Principal Component Extraction", *IEEE Transactions on Neural Networks*, Vol. 9, No. 1, pp. 58-67, Jan 1998.

[84]. T. Chen, S. I. Amari, and Q. Lin, "A unified algorithm for principal and minor components extraction", *Neural Networks*, Vol. 11, pp. 382-390, 1998.

[85]. D. Z. Feng, Z. Bao, and L. C. Jiao, "Total Least Mean Squares Algorithm", *IEEE Transactions on Signal Processing*, Vol. 46, No. 8, pp. 2122-2130, Aug 1998.

[86]. Y. Miao and Y. Hua, "Fast Subspace Tracking and Neural Network Learning by a Novel Information Criterion", *IEEE Trans. on Signal Proc.*, Vol. 46, No. 7, pp. 1967-1979, 1998.

[87]. P. Strobach, "Fast Orthogonal Iteration Adaptive Algorithms for the Generalized for Symmetric Eigenproblem", *IEEE Trans. on Signal Processing*, Vol. 46, No. 12, 1998.

[88]. J-P. Delmas and J. F. Cardoso, "Asymptotic Distributions Associated to Oja's Learning Equation for Neural Networks", *IEEE Trans. on Neural Networks*, Vol. 9, No. 6, 1998.

[89]. J-P. Delmas and J. F. Cardoso, "Performance Analysis of an Adaptive Algorithm for Tracking Dominant Subspaces", *IEEE Trans. on Signal Proc.*, Vol. 46, No. 11, pp. 3045-3057, 1998.

[90]. J-P. Delmas and F. Alberge, "Asymptotic Performance Analysis of Subspace Adaptive Algorithms Introduced in the Neural Network Literature", *IEEE Trans. on Signal Processing*, Vol. 46, No. 1, pp. 170-182, 1998.

[91]. S. C. Douglas, S. Y. Kung, and S. Amari, "A self-stabilized minor subspace rule", *IEEE Signal Processing Letters*, 5, pp. 328-330, 1998.

[92]. V. Solo, "Performance Analysis of Adaptive Eigenanalysis Algorithms", *IEEE Trans. on Signal Processing*, Vol. 46, No. 3, pp. 636-646, 1998.

[93]. P. Strobach, "Fast orthogonal Iteration Adaptive Algorithms for Generalized Symmetric Eigenproblem", *IEEE Trans. on Signal Processing*, Vol. 46, No. 12, pp. 3345-3359, 1998.

[94]. C. Chatterjee, Z. Kang, and V. P. Roychowdhury, "Adaptive Algorithms for Accelerated PCA from an Augmented Lagrangian Cost Function", *Proc. Int'l Joint Conference on Neural Networks (IJCNN '99)*, July 10-16, Washington D.C., pp. 1043-1048, Vol 2, 1999.

[95]. A. Taleb and G. Cirrincione, "Against the Convergence of the Minor Component Analysis Neurons", *IEEE Transactions on Neural Networks*, Vol. 10, No. 1, pp. 207-210, Jan 1999.

[96]. A. R. Webb, "A loss function approach to model selection in nonlinear principal components", *Neural Networks*, 12, 339-345, 1999.

[97]. S. Ouyang, Z. Bao, and G. Liao, "A class of learning algorithms for principal component analysis and minor component analysis", *Electronics Letters*, 35, pp. 443-444, 1999.

[98]. F. L. Luo and R. Unbehauen, "Comments on: A unified algorithm for principal and minor component extraction", *Neural Networks*, 12, 1999.

[99]. C. Chatterjee, Z. Kang, and V. P. Roychowdhury, "Algorithms For Accelerated Convergence Of Adaptive PCA", *IEEE Trans. on Neural Networks*, Vol. 11, No. 2, pp. 338-355, March 2000.

[100]. Y-F. Chen, M. D. Zoltowski, J. Ramos, C. Chatterjee, and V. Roychowdhury, "Reduced Dimension Blind Space-Time 2-D RAKE Receivers for DS-CDMA Communication Systems", *IEEE Trans. on Signal Processing*, Vol. 48, No. 6, pp. 1521-1536, June 2000.

[101]. Q. Zhang and Y-W. Leung, "A Class of Learning Algorithms for Principal Component Analysis and Minor Component Analysis", *IEEE Transactions on Neural Networks*, Vol. 11, No. 2, pp. 529-533, March 2000.

[102]. Z. Kang, C. Chatterjee, and V. P. Roychowdhury, "An Adaptive Quasi-Newton Algorithm for Eigensubspace Estimation", *IEEE Transactions on Signal Processing*, Vol. 48, No. 12, pp. 3328-3335, December 2000.

[103]. S. Ouyang, Z. Bao, and G-S. Liao, "Robust Recursive Least Squares Learning Algorithm for Principal Component Analysis", *IEEE Trans. On Neural Networks*, Vol. 11, No. 1, 2000.

[104]. A. Weingessel and K. Hornik, "Local PCA Algorithms", *IEEE Transactions on Neural Networks*, Vol. 11, No. 6, 2000.

[105]. R. Moller, "A Self-Stabilizing Learning Rule for Minor Component Analysis", *Int'l Journ. Of Neural Systems*, April 2000.

[106]. S. Ouyang, Z. Bao, G. S. Liao, and P. C. Ching, "Adaptive Minor Component Extraction with Modular Structure", *IEEE Trans. Signal Proc.*, Vol. 49, No. 9, pp. 2127-2137, 2001.

[107]. D. Feng, Z. Bao, and X-D. Zhang, "A Cross-Associative Neural Network for SVD of Nonsquared Data Matrix in Signal Processing", *IEEE Transactions On Neural Networks*, Vol. 12, No. 5, 2001.

REFERENCES

[108]. T. Chen and S. Amari, "Unified stabilization approach to principal and minor component extraction", *Neural Networks*, 14, pp. 1377-1387, 2001.

[109]. G. Cirrincione, M. Cirrincione, J. Herault, and S. VanHuffel, "The MCA EXIN Neuron for Minor Component Analysis", *IEEE Trans. on Neural Networks*, Vol. 13, No. 1, pp. 160-187, Jan 2002.

[110]. S. Ouyang and Z. Bao, "Fast Principal Component Extraction by a Weighted Information Criterion", *IEEE Trans. Signal Processing*, Vol. 50, No. 8, pp. 1994-2002, August 2002.

[111]. P. J. Zufiria, "On the Discrete-Time Dynamics of the Basic Hebbian Neural Network Node", *IEEE Trans. Neural Networks*, Vol. 13, No. 6, 2002.

[112]. J. A. K. Suykens, T. Van Gestel, J. Vandewalle, and B. DeMoor, "A Support Vector Machine Formulation to PCA Analysis and Its Kernel Version", *IEEE Trans. Neural Networks*, Vol. 14, No. 2, 2003.

[113]. S. Ouyang, P. C. Ching, and T. Lee, "Robust adaptive quasi-Newton algorithms for eigensubspace estimation", *IEEE Proc. Vision Image and Signal Processing*, Vol. 150, No. 5, pp. 321-330, 2003.

[114]. R. Moller and A. Konies, "Coupled Principal Component Analysis", *IEEE Trans. Neural Networks*, Vol. 15, No. 1, 2004.

[115]. D-Z. Feng, W-X. Zheng, and Y. Jia, "Neural Network Learning Algorithms for Tracking Minor Subspace in High-Dimensional Data Stream", *IEEE Trans. Neural Networks*, Vol. 16, No. 3, 2005.

[116]. Z. Yi, M. Ye, J.C. Lv, and K. K. Tan, "Convergence Analysis of a Deterministic Discrete Time System of Oja's PCA Learning Algorithm", *IEEE Trans. Neural Networks*, Vol. 16, No. 6, 2005.

[117]. C. Chatterjee, "Adaptive Algorithms for First Principal Eigenvector Computation", *Neural Networks*, Vol. 18, No. 2, pp. 145-149, March 2005.

[118]. M. V. Jankovic and H. Ogawa, "Modulated Hebb-Oja Learning Rule- A Method for Principal Subspace Analysis", *IEEE Trans. Neural Networks*, Vol. 17, No. 2, 2006.

[119]. M. Ye, X-Q. Fan, and X. Li, "A Class of Self-Stabilizing MCA Learning Algorithms", *IEEE Trans. Neural Networks*, Vol. 17, No. 6, 2006.

[120]. K. A. Brakke, J. M. Mantock, and K. Fukunaga, "Systematic Feature Extraction", *IEEE Transactions on Pattern Analysis and Machine Vision*, Vol. 4, No. 3, pp. 291-297, 1982.

[121]. M. Artin, *Algebra*, Englewood Cliffs, NJ: Prentice Hall, 1991.

[122]. B. D. O. Anderson and J. B. Moore, *Optimal Control - Linear Quadratic Methods*, Prentice Hall, New Jersey, 1990.

[123]. A. Benveniste, A. Metivier, and P. Priouret, *Adaptive Algorithms and Stochastic Approximations*, New York: Springer-Verlag, 1990.

[124]. P. J. Bickel and K. A. Doksum, *Mathematical Statistics*, Holden-Day Inc., Oakland, CA, 1977.

[125]. G. Birkhoff and G-C. Rota, *Ordinary Differential Equations*, Second Edition, Blaisdell Publishing Co., Massachusetts, 1969.

[126]. K. A. Brakke, J. M. Mantock, and K. Fukunaga, "Systematic Feature Extraction", *IEEE Transactions on Pattern Analysis and Machine Vision*, Vol. 4, No. 3, pp. 291-297, 1982.

[127]. C. Chatterjee and V. P. Roychowdhury, "A New Training Rule for Optical Recognition of Binary Character Images by Spatial Correlation", *Proceedings IEEE Int'l Conference on Neural Networks (ICNN '94)*, June 28-July 2, 1994, Orlando, Florida, pp. 4095-4100.

[128]. A. Cichocki and R. Unbehauen, "Neural networks for solving systems of linear equations and related problems", *IEEE Trans. Circuits Syst.*, Vol. 39, pp. 124-198, 1992.

[129]. A. Cichocki and R. Unbehauen, "Simplified Neural Networks for Solving Linear Least Squares Problems in Real Time", *IEEE Trans. Neural Networks*, Vol. 5, No. 6, pp. 910-923, 1994.

[130]. A. Cichocki and R. Unbehauen, *Neural Networks for Optimization and Signal Processing*, John Wiley and Sons, New York, 1993.

[131]. D. M. Clark and K. Ravishankar, "A Convergence Theorem for Grossberg Learning", *Neural Networks*, Vol. 3, pp. 87-92, 1990.

[132]. P. A. Devijver, "Relationship between Statistical Risks and the Least-Mean-Square Error Design Criterion in Pattern Recognition", *First Int'l Joint Conf. on Patt. Recog.*, Washington D.C., pp. 139-148, 1973.

[133]. P. A. Devijver, "On a New Class of Bounds on Bayes Risk in Multihypothesis Pattern Recognition", *IEEE Trans. on Computers*, Vol. C-23, No. 1, pp. 70-80, 1974.

[134]. P. A. Devijver and J. Kittler, *Pattern Recognition: A Statistical Approach*, Prentice Hall International, Englewood Cliffs, NJ, 1982.

[135]. R. O. Duda and P. E. Hart, *Pattern Classification and Scene Analysis*, John Wiley and Sons, New York, 1973.

[136]. D. H. Foley and J. W. Sammon, "An Optimal Set of Discriminant Vectors", *IEEE Transactions on Computers*, Vol.c-24, No. 3, pp. 281-289, March 1975.

[137]. K. Fukunaga, *Introduction to Statistical Pattern Recognition*, Second Edition, Academic Press, New York, 1990. www.amazon.com/Introduction-Statistical-Recognition-Scientific-Computing/dp/0122698517.

[138]. K. Fukunaga and W. L. G. Koontz, "Application of the Karhunen-Loeve expansion to feature selection and ordering", *IEEE Trans. Comput.*, Vol. C-19, pp. 311-318, 1970.

[139]. P. Gallinari, S. Thiria, F. Badran, and F. Fogelman-Soulie, "On the Relations Between Discriminant Analysis and Multilayer Perceptrons", *Neural Networks*, Vol. 4, pp. 349-360, 1991.

[140]. H. Gish, "A probabilistic approach to the understanding and training of neural network classifiers", in *Proc. IEEE Conf. on Acoust. Speech and Signal Proc.*, pp. 1361-1364, 1990.

[141]. G. H. Golub and C. F. VanLoan, *Matrix Computations*, Baltimore, MD: Johns Hopkins Univ. Press, 1983.

[142]. J. B. Hampshire II and B. Pearlmutter, "Equivalence Proofs for Multi-Layer Perceptron Classifiers and the Bayesian Discriminant Function", *Connectionist Models - Proc. of the 1990 Summer School*, Ed. D.S.Touretzky et al., pp. 159-172, 1990.

[143]. S. Haykin, *Neural Networks - A Comprehensive Foundation*, Maxwell Macmillan International, New York, 1994.

[144]. J. Hertz, A. Krogh, and R. G. Palmer, *Introduction to the Theory of Neural Computation*, Addison-Wesley Publishing Co., California, 1991.

[145]. M. Honig, U. Madhow, and S. Verdu, "Blind Adaptive Multiuser Detection", *IEEE Transactions on Information Theory*, Vol. 41, No. 4, pp. 944-960, July 1995.

[146]. J. J. Hopfield, "Neurons with graded response have collective computational properties like those for two-state neurons", *Proc. Natl. Acad. Science USA*, Vol. 81, pp. 3088-3092, 1984.

[147]. K. Hornik and C-M. Kuan, "Convergence Analysis of Local Feature Extraction Algorithms", *Neural Networks*, Vol. 5, pp. 229-240, 1989.

[148]. A. K. Jain and J. Mao, "Artificial Neural Network for Nonlinear Projection of Multivariate Data", *Proc. IJCNN*, Vol. 3, Baltimore, Maryland, June 1992.

[149]. A. K. Jain and J. Mao, "Artificial Neural Network for Feature Extraction and Multivariate Data Projection", *IEEE Trans. Neural Networks*, Vol. 6, pp. 296-316, 1995.

[150]. J. Karhunen and J. Joutsensalo, "Representation and Separation of Signals Using Nonlinear PCA Type Learning", *Neural Networks*, Vol. 7, No. 1, pp. 113-127, 1994.

[151]. T. Kohonen, *Self-Organization and Associative Memory*, Springer-Verlag, Berlin, 1984.

[152]. E. Kreyszig, *Advanced Engineering Mathematics*, 6th edition, Wiley, New York 1988.

[153]. S. Y. Kung, *Digital Neural Networks*, Englewood Cliffs, NJ: Prentice Hall, 1992.

[154]. H. J. Kushner and D. S. Clark, Stochastic Approximation Methods for Constrained and Unconstrained Systems, Springer-Verlag, New York, 1978.

[155]. D. Le Gall, "MPEG: A video compression standard for multimedia applications", *Commns. of the ACM*, Vol. 34, pp. 46-58, 1991.

[156]. R. P. Lippman, "An introduction to computing with neural nets", *IEEE ASSP Magazine*, pp. 4-22, 1987.

[157]. L. Ljung, "Analysis of Recursive Stochastic Algorithms", *IEEE Transactions on Automatic Control*, Vol. AC-22, No. 4, pp. 551-575, August 1977.

[158]. L. Ljung, "Strong Convergence of a Stochastic Approximation Algorithm", *The Annals of Statistics*, Vol. 6, No. 3, pp. 680-696, 1978.

[159]. L. Ljung, "Analysis of Stochastic Gradient Algorithms for Linear Regression Problems", *IEEE Transactions on Information Theory*, Vol. 30, No. 2, pp. 151-160, 1984.

[160]. L. Ljung, G. Pflug, and H. Walk, *Stochastic Approximation and Optimization of Random Systems*, Boston: Birkhauser Verlag, 1992.

[161]. D. G. Luenberger, *Linear and Nonlinear Programming*, Second Edition, Addison-Wesley Publishing Company, Reading Massachussets, 1984.

[162]. F. Luo and R. Unbehauen, *Applied Neural Networks for Signal Processing*, Cambridge U.K., Cambridge Univ. Press, 1997.

[163]. R. Lupas and S. Verdu, "Near-far resistance of multi-user detectors in asynchronous channels", *IEEE Transactions on Communications*, Vol. 38, pp. 496-508, Apr. 1990.

[164]. U. Madhow and M. L. Honig, "MMSE Interference Suppression for Direct-Sequence Spread-Spectrum CDMA", *IEEE Trans. on Communications*, Vol. 42, No. 12, pp. 3178-3188, 1994.

[165]. S. Miyake and F. Kanaya, "A Neural Network Approach to a Bayesian Statistical Decision Problem", *IEEE Trans. on Neural Networks*, Vol. 2, No. 5, pp. 538-540, 1991.

[166]. B. K. Moor, "ART 1 and pattern clustering", *Proc. 1988 Connectionist Summer School*, pp. 174-185, Morgan-Kaufman, 1988.

[167]. L. Niles, H. Silverman, G. Tajchman, and M. Bush, "How limited training data can allow a neural network to outperform an optimal statistical classifier", *Proc. of the ICASSP*, pp. 17-20, 1989.

[168]. T. Okada and S. Tomita, "An Optimal Orthonormal System for Discriminant Analysis", *Pattern Recognition*, Vol. 18, No. 2, pp. 139-144, 1985.

[169]. N. Otsu, "Optimal linear and nonlinear solutions for least-square discriminant feature extraction", *Proc. 6th Int'l Conf. on Patt. Recog.*, Vol. 1, Germany, pp. 557-560, 1982.

[170]. N. L. Owsley, "Adaptive data orthogonalization", Proc. 1978 IEEE Int. Conf. on Acoustics, Speech, and Signal Processing, pp. 109-112, 1978.

[171]. N. R. Pal, J. C. Bezdek, and E. C-K. Tsao, "Generalized clustering networks and Kohonen's self-organizing scheme", *IEEE Trans. Neural Networks*, Vol. 4, No. 4, pp. 549-557, 1993.

[172]. J. D. Patterson, T. J. Wagner, and B. F. Womack, "A Mean-Square Performance Criterion for Adaptive Pattern Classification Systems", *IEEE Transactions on Automatic Control*, Vol.12, pp. 195-197, 1967.

[173]. V. F. Pisarenko, "The retrieval of harmonics from a covariance function", *Geophysics Journal of Royal Astronomical Society*, Vol. 33, pp. 347-366, 1973.

[174]. Y. Hua and T. K. Sarkar, "On SVD for Estimating Generalized Eigenvalues of Singular Matrix Pencil in Noise", *IEEE Trans. on Signal Processing*, Vol. 39, No. 4, pp. 892-900, 1991

[175]. M. D. Richard and R. P. Lippmann, "Neural Network Classifiers Estimate Bayesian a posteriori Probabilities", *Neural Computation*, Vol. 3, pp. 461-483, 1991.

[176]. D. A. Robinson, "The use of control systems analysis in the neurophysiology of eye movement", *Annual Review of Neuroscience*, Vol. 4, pp. 463-503, 1981.

[177]. D. W. Ruck, S. K. Rogers, M. Kabrisky, M. E. Oxley, and B. W. Suter, "The Multilayer Perceptron as an Approximation to a Bayes Optimal Discriminant Function", *IEEE Transactions on Neural Networks*, Vol.1, No.4, pp.296-298, 1990.

[178]. D. E. Rumelhart and J. L. McClelland, *Parallel and Distributed Processing*, The MIT Press, Cambridge, MA, 1986.

[179]. S. Verdu, "Multiuser detection", *Advances in Detection and Estimation*, JAI Press, 1993.

[180]. E. A. Wan, "Neural Network Classification: A Bayesian Interpretation", *IEEE Trans. on Neural Networks*, Vol. 1, No. 4, pp. 303-305, Dec. 1990.

[181]. A. R. Webb and D. Lowe, "The Optimised Internal Representation of Multilayer Classifier Networks Performs Nonlinear Discriminant Analysis", *Neural Networks*, Vol. 3, pp. 367-375, 1990.

[182]. R. L. Wheeden and A. Zygmund, *Measure and Integral - An Introduction to Real Analysis*, Marcel Dekker, Inc., New York, 1977.

[183]. R. H. White, "Competitive Hebbian Learning: Algorithm and Demonstrations", *Neural Networks*, Vol. 5, pp. 261-275, 1992.

[184]. R. J. Williams, "Feature discovery through error-correction learning", Institute of Cognitive Science, Univ. of California, San Diego, Tech. Rep. 8501, 1985.

[185]. H-C. Yau and M. T. Manry, "Iterative Improvement of a Gaussian Classifier", *Neural Networks*, Vol. 3, pp. 437-443, 1990.

[186]. F. McNamee et al. "A Case For Adaptive Deep Neural Networks in Edge Computing", December 2016.

[187]. Vinicius M. A. Souza et al., :Challenges in Benchmarking Stream Learning Algorithms with Real-world Data, Journal Data Mining and Knowledge Discovery," Apr 2020. https://arxiv.org/pdf/2005.00113.pdf.

[188]. Publicly real-world datasets to evaluate stream learning algorithms, https://sites.google.com/view/uspdsrepository.

REFERENCES

[189]. M. Apczynski et al. (2013), "Discovering Patterns of Users' Behaviour in an E-shop - Comparison of Consumer Buying Behaviours in Poland and Other European Countries", "Studia Ekonomiczne", nr 151 p. 144-153.

[190]. Stratus Technologies, "Gartner 2021 Strategic Roadmap for Edge Computing", Johannesburg, 07 Jun 2021, www.itweb.co.za/content/lwrKx73KXLL7mg1o.

[191]. Kaz Sato et al., "Monitor models for training-serving skew with Vertex AI", 2021. https://cloud.google.com/blog/topics/developers-practitioners/monitor-models-training-serving-skew-vertex-ai.

[192]. Christoph H. Lampert, et al., Printing Technique Classification for Document Counterfeit Detection, 2006 International Conference on Computational Intelligence and Security, Nov 2006.

[193]. Yahoo Research Webscope Computer Systems Data. S5 - A Labeled Anomaly Detection Dataset, version 1.0 (16M). https://webscope.sandbox.yahoo.com/catalog.php?datatype=s&did=70.

[194]. Shay Palachy, Detecting stationarity in time series data. Towards Data Science, 2019. https://towardsdatascience.com/detecting-stationarity-in-time-series-data-d29e0a21e638.

[195]. Chanchal Chatterjee Github: https://github.com/cchatterj0/AdaptiveMLAlgorithms.

[196]. Standard basis, Wikipedia, https://en.wikipedia.org/wiki/Standard_basis.

[197]. Keras, MNIST digits classification dataset. https://keras.io/api/datasets/mnist/.

[198]. Neuromorphic engineering, Wikipedia. https://en.wikipedia.org/wiki/Neuromorphic_engineering.

[199]. Karhunen-Loève theorem, Wikipedia. https://en.wikipedia.org/wiki/Karhunen%E2%80%93Lo%C3%A8ve_theorem.

[200]. Principal component analysis, Wikipedia. https://en.wikipedia.org/wiki/Principal_component_analysis.

[201]. Linear discriminant analysis, Wikipedia. https://en.wikipedia.org/wiki/Linear_discriminant_analysis.

[202]. Generalized eigenvector, Wikipedia. https//en.wikipedia.org/wiki/Generalized_eigenvector.

[203]. Singular value decomposition, Wikipedia. https://en.wikipedia.org/wiki/Singular_value_decomposition.

[204]. Autoencoder, Wikipedia. http://en.wikipedia.org/wiki/Autoencoder.

[205]. Sherman-Morrison formula, Wikipedia. https://en.wikipedia.org/wiki/Sherman%E2%80%93Morrison_formula.

REFERENCES

[206]. Linear discriminant analysis, Wikipedia.
https://en.wikipedia.org/wiki/Linear_
discriminant_analysis.

[207]. Cholesky decomposition, Wikipedia.
https://en.wikipedia.org/wiki/Cholesky_
decomposition.

[208]. Frobenius norm, Wikipedia. https://
en.wikipedia.org/wiki/Matrix_
norm#Frobenius_norm.

[209]. Nonlinear conjugate gradient method, Wikipedia.
https://en.wikipedia.org/wiki/Nonlinear_
conjugate_gradient_method.

Index

A, B

Printed in the United States
by Baker & Taylor Publisher Services